SCHOLASTIC

create and display

Science

Full of exciting activities and displays for the whole curriculum

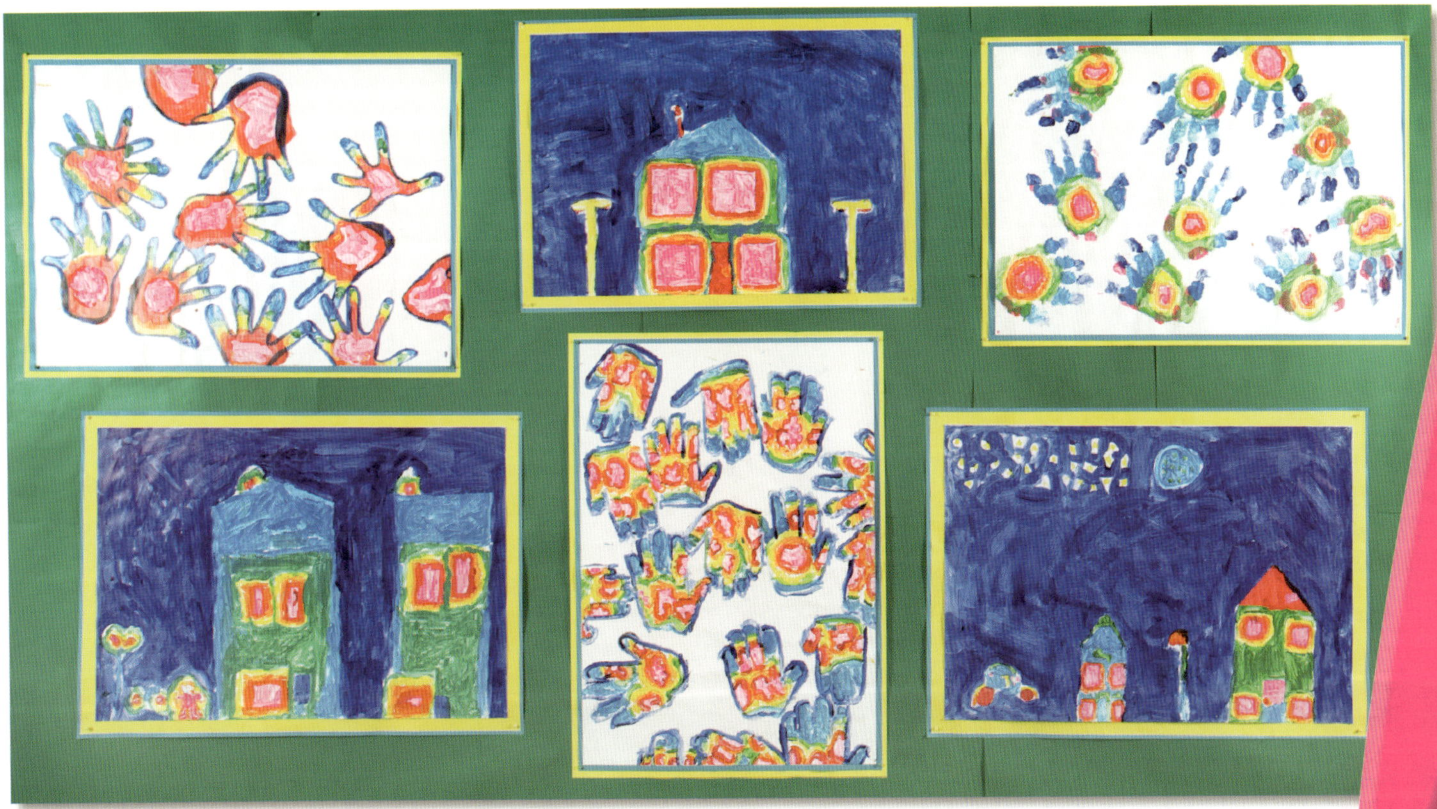

Ages 5-11

Rebecca Carnihan

SCHOLASTIC

Book End, Range Road, Witney, Oxfordshire, OX29 OYD

www.scholastic.co.uk

© 2012, Scholastic Ltd

1 2 3 4 5 6 7 8 9   0 1 2 3 4 5 6 7 8 9

British Library Cataloguing-in-Publication Data
A catalogue record for this book is available from the British Library.

ISBN 978-1407-12529-9
Printed by Bell & Bain Ltd, Glasgow

Text © 2012 Rebecca Carnihan

Commissioning Editor
Paul Naish

Development Editor
Emily Jefferson

Editor
Kate Greig

Series Designers and Cover Design
Andrea Lewis and Sarah Garbett

Photography:
Gareth Boden and Steve Forrest

Acknowledgements

The author and publisher would like to thank Executive Headteacher Mrs Cynthia Eubank, Head of School Ms Carol Wilson, all staff and especially the children of Grinling Gibbons Primary School in Deptford, London, for their hard work, enthusiasm and pride in the production of work for this book.

A special thank you is given to Lucy Pashley who gets the very best from the infant classes and whose work was beautifully executed. And to my mum, Hilary Carnihan, who tirelessly supports my creative ideas and endeavours. Thank you.

Images

Page 53, *Puppy* sculpture by artist Jeff Koons © Tom Sanchez Poy/Shutterstock

Contents

Introduction

This book contains a wide range of art activities and display ideas based on primary Science topics. There are five chapters covering different aspects of Science, including: Chemistry, Physics, Technology, Human Biology, Plant Biology, Astronomy and Geology.

Projects for the youngest children are generally at the beginning of each chapter, though there are many overlapping themes that can be adapted for different age groups.

The aim is for these art activities to deepen the children's understanding of Science, either by making art that uses similar processes to those in any given Science project, or through the representation of a scientific idea.

There are a wide variety of media and techniques, with 2D and 3D work throughout the book. Many projects use core resources, including: PVA glue, glue sticks, sheets of flexible card, acrylic paints, printing inks, a photocopier, cartridge paper, black sugar paper and cardboard boxes (it is useful to collect, flatten and store these after school deliveries).

It is a good idea to build up a box of resources, materials and clean recyclables that have interesting textures, such as string, plastic

containers, wool, sandpaper, fabric scraps, twigs, buttons, bottle lids, lolly sticks, corks, tissue paper, cotton wool, sponges, paper art straws, raffia and pipe cleaners. These can be used to make impressions, textures or incorporated into a finished artwork.

For interesting resources, find out if there is a children's scrap project in your local area – they are usually charitable organisations that gather clean waste materials from manufacturers specifically for this use. Or contact your local council for information about similar schemes.

I hope that these activities prove exciting and bring out the best in your children. There is scope for them to contribute to and innovate within these projects and you will find the children's contribution to a project's development leads to a valuable and stimulating exchange of ideas.

Rebecca Carnihan

Paper Play

This is a good project for young children to improve their motor skills using scissors, and is an architectural project in its own right. The children will identify the forms – including a bridge, roller coaster, river, playground – in imaginative ways.

Resources

- A variety of coloured paper
- Stiff board or mount board, approximately A5
- Glue sticks
- Scissors
- Drawing materials

Approach

1 Hand out a variety of coloured paper and demonstrate cutting strips approximately 2cm wide. The aim for the child is to cut slowly and with care to produce even strips.
2 Take the first strip and spread glue on one end and stick to the board. Carefully lift the other end and form a 'bridge' or loop and glue the other end in position.
3 Continue adding more strips, twisting over and under the previous strips, to form a series of colourful loops.

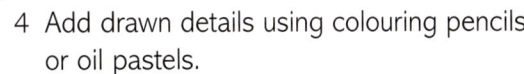

4 Add drawn details using colouring pencils or oil pastels.

Plastic Fantastic

In the 1960s plastic was a popular 'modern' material. It was used by designers to make furniture, lampshades, household products, raincoats, boots and even clothes. The change in technology enabled designers to produce plastic products in any colour, adding fun to everyday objects. This project uses simple but structured materials to make a colourful sculpture.

Resources

- Plastic disposable cups
- Acrylic paint
- Flexible white, black, coloured or patterned card
- Scissors
- Stapler (teacher use only)
- Sticky tape

Approach

1 Paint several plastics cups in different colours – both inside and out – changing the colour combination each time. When the outside of each cup is dry, paint a bright or metallic colour onto the rim, or dip into paint to achieve the same effect. Put aside to dry completely.

2 For the stand cut a piece of card approximately 60cm in width and to any height you like.

3 Draw around the top of a cup several times in different places on the card. Poke a hole in the centre of each circle using a pencil (teacher to do this).

4 Continue by cutting lines from the hole in the centre of the circle to the edge of the circle, until it resembles the spokes of a wheel.

5 Bend the card structure into a cylinder and staple at both ends and use sticky tape on the inside to join the seam. This will now stand upright on the table.

6 Gently bend back the segments in each circle and push a cup inside. The segments should be bent back enough to hold each cup securely in position. Some should be pushed in more than others for a variety of protrusions.

7 Once all the cups are in place the work is complete, and can stand on a flat surface or be hung from the ceiling with cotton.

Bending and Squeezing

This is an exploratory project for children to play with and use materials in the following ways – bending, twisting, stretching, pressing and wrapping. Some materials will be plastic, some elastic, some flexible and others not. It is all about discovery, arriving at a final object through the journey of play.

Resources

- Plasticine®
- Wrapping materials, for example, string, pipe-cleaners, ribbons, insulation tape
- Modelling wire
- Buttons

Approach

1 Take a piece of Plasticine®, manipulate it by bending and squeezing but without pulling it apart. Leave in an amorphous shape (with young children you can shout 'stop' so that they leave a spontaneous form).

2 Using a selection of the wrapping materials wrap, tie and knot together the materials to cover the whole of the Plasticine® shape. Explore and have fun with the materials available.

3 After this process ask the group to think about what their form could be, and to continue modelling it with this in mind. As desired, add buttons using the modelling wire. Poke it through the button holes and into the Plasticine® together with any of the other materials available.

A Golden Sun

This project looks at the effects of heat on bread, and uses the toast to make a picture! The Sun is a suitable focus of study for this activity as it is a source of warmth and has a simple stylised shape. Start with a circular centre and create any shape and size of sun rays you like. You can use this technique to create another image, such as flowers or a street scene.

Resources

- Thin or medium sliced bread
- A toaster
- Scissors
- Black card or sugar paper
- PVA glue

Approach

1 To prepare, toast some of the bread to a light shade, and toast some until it is golden brown. This will give you several shades to work with.
2 Use scissors to cut the toast into shapes or to cut lines from the crusts.
3 Glue each piece of toast onto the black card using generous amounts of PVA glue and allow to dry.
4 As the work will not last indefinitely it should be photographed. Check the work for signs of mould, upon which it should be thrown away.

Stretching Installation

Ernesto Neto (1964–) is a contemporary artist who creates large-scale installations with hanging structures made from stocking-like material. Show the children images of his work and discuss the use of space – such as galleries or churches – and gravity. The group piece that you make will be a 'mini' installation and the weights can be made from fragranced materials, light and heavy materials, bulky materials or a combination.

Resources

For the sculptures
- Pop socks
- Acrylic paint in a variety of colours
- Rice
- Sand
- Pulses
- Buttons

For the structure
- A large cardboard box
- A large sheet of paper or flexible card
- PVA glue
- Three or four bamboo sticks
- A digital camera
- Camomile
- Lavender

Approach

1 Dye the pop socks in the acrylic paint colour of your choice with a little water and hang to dry.
2 Remove the top and one side of the box and tape the remaining sides to secure its shape.
3 Cut the paper to the same height as the box and place carefully as a panoramic curve within the box and secure with tape at the edges.
4 Paint the bamboo sticks white to match the background (optional) and poke through the top of the box (from side to side) to create a criss-cross structure.
5 Ask each child to choose a filling and place some in their coloured sock. Suspend each sock by tying it to the bamboo sticks.
6 In the meantime photograph the children individually posing looking up in surprise, wonder or curiosity. Print the images so they are approximately 15cm in height.
7 Place most of the figures on the panoramic curve. Some of the figures can be free-standing, stick them onto flexible card with a flap at the bottom to bend, glue and place in position.

Stretching Geometry

This project uses simple materials to create free-flowing yet geometrical forms. The placing of the pins has no order and the thread stretched around each point can be taken in any direction like a dot-to-dot drawing. As a starting point look at the work of artists Joan Miró (1893–1983), Paul Klee (1879–1940) and Wassily Kandinsky (1866–1944) whose work is very playful and colourful. Made on a large scale it makes an interesting surround for a maths display.

Resources

- Map pins
- Colourful threads
- Thick cardboard
- Coloured paper
- Glue sticks

Approach

1 Cut the cardboard into pieces, for example, 40cm x 20cm or 30cm x 30cm.
2 For safety reasons the card must remain on the table whilst working. Place the pins randomly on the card, spacing them out.
3 Start with one colour thread and tie around the first pin (the children may need a little help) then move onto a neighbouring pin, wind the thread around it and continue to the next pin, wrapping around it once to secure the thread. Continue until the thread runs out and secure with a knot.
4 Choose another colour and create a 'mapping' pattern by repeating the above process.

5 Once complete add small pieces of coloured paper to the shapes created.
6 The completed pieces should be displayed on a wall (they should be handled by the teacher).

Thermal Paintings

Technology can show us degrees of heat through the use of thermal imaging (images can be found on the internet). Discuss with the children things they think give off heat, for example, the human body, radiators, occupied and unoccupied buildings. Explain how useful thermal readers are to detect people trapped in collapsed buildings, and for measuring how much heat is being lost from a building or saved by good insulation. This painting project looks at the heat coming from a building. Alternatively, you can look at body heat – ask the children to feel for the warmest part of the hand (the centre of the palm), and which parts of their hands feel coldest when coming in after playtime (the fingertips). These variations in temperature can be illustrated with a thermal image painting.

Resources

- Acrylic or poster paint in the following colours – hottest to coolest: pink, red, orange, yellow, green, light blue, dark blue
- A2 or A3 paper

Approach

1 Place a hand on the paper and draw around it several times, changing direction each time.
2 To show thermal heat and insulation in buildings, draw a simple street scene with a house, perhaps adding a car, street lamps, or people.
3 Starting with the warmest parts of the picture use pink to paint the centre of the hand as it is the warmest part, and the house windows as they release the most heat.
4 Proceed by using the increasingly cool colours, use red to surround the pink area, orange next to the red, yellow next to the orange, green next to the yellow, light blue next to the green and finally dark blue next to the light blue.

Keeping Warm – Funky Sheep

Wool is one of the best insulators to wear to keep warm, and this project shows wool's journey from sheep to textile. A good starting point is to look at images of sheep with long coats, the shearing process, dyeing or dyed fleeces, the spinning stage, wool in skeins, a ball of wool, and its destination as knitted clothing, blankets and tapestries, etc. Textile designer Kaffe Fassett (1937–) produces very colourful knitted work, images of which can be found on the internet.

Resources

- Stiff card
- Scissors
- Wool in a variety of colours
- Black crayon or paint
- Buttons
- PVA glue and/or glue gun (teacher use only)

Approach

1 Draw a template of a sheep to the desired size, the sheep pictured here are 45cm x 30cm, but can be smaller depending on how much display space or wool is available.

2 Draw around the template on a piece of card and cut it out. Children may need help with this. Alternatively, the teacher should cut out each sheep using a cutting knife and mat if the card is very stiff.

3 Select three strands of wool in the colour of the children's choice in as long a length as is manageable to work with.

4 Start by tying the ends of the three strands together. Next tie them in a double knot around the belly of the sheep. Then begin to wrap around the sheep, building up the wool width-wise. It can be wrapped in any direction, but needs to avoid covering the legs and head.

5 Each time the threads come to an end tie a knot at the back, and repeat the process again (see point 4). Thread colours can be altered a little to add some variety to the sheep as the 'funky fleece' builds up.

6 Once the wrapping is complete (either fully or partially as desired) go on to create some interesting details. Create these by pinching and tying strands together, or using the same colours to make a loop and weave it in.

7 Use a button for an eye. Use PVA glue to stick the button in place and lay flat to dry. Use black paper to cut shapes for the facial features and hooves.

8 For a 'standing' sheep use the card off-cuts in a right-angled triangle shape and glue to the back of the legs with the glue gun (teacher to do this). Alternatively, the work can be pinned to a display board as a two-dimensional piece.

Paper-Pulp Sculptures

These starfish are a colourful shape made from paper-pulp. The project can be adapted for other shapes, as long as the Plasticine® can be easily removed once the paper-pulp has hardened. Paper-pulp is surprisingly tough and demonstrates that liquids and solids can be combined to transform one another, and create new forms in the process once evaporation has occurred.

Resources

- Plasticine®
- Tissue paper in a variety of colours (1 sheet per colour)
- Metallic paper
- PVA glue
- Large plastic bowl
- A tray or piece of cardboard
- Cling wrap

Approach

1 Make a starfish shape with the Plasticine®. Put a piece of cling wrap on the tray or card and place the Plasticine® model on top.

2 Mix half a cup of PVA glue with a little water and pour into the plastic bowl.

3 Choose a maximum of four colours. Good starfish colours are yellow, orange and red, though they can be any colour combination.

Tear the paper into pieces and add to the plastic bowl, along with some pieces of metallic paper, turning them over so that they are soaked.

4 Once the paper is soaking wet start to tear the paper into small pieces, this will take a little while; the smaller the pieces the better the pulp mixture will be.

5 When the paper and water have formed a pulp begin to cover the Plasticine® starfish on top and over the sides, but not underneath.

6 Allow to dry, turn the starfish over and carefully scoop out the Plasticine® leaving the remaining starfish sculpture. This is easier to do in a warm place as the Plasticine® will soften.

Salty Seascapes

To create the foaming crest of a wave in these collages, water (liquid) mixed with salt (solid) are used with the process of evaporation. Some starting points to this project include looking at the process of salt harvesting at salt farms. Look at a selection of ocean photos that show the movement of the sea, shapes of the waves and foaming crests. For an artist's impression of the sea, look at the seascapes of JMW Turner (1775–1851).

Resources

- Table salt
- Blue sugar paper in a variety of shades
- Silver paper
- A2 cartridge paper
- Glue sticks

Approach

1 Tear the blue paper into strips and assemble them on the cartridge paper, glue into position once happy with the formations made.

2 Stir four tablespoons of salt into half a standard plastic art beaker of warm water. Not all of it will dissolve, but it will retain crystals that will glisten on the finished work.

3 Tear strips which curl up and add some three-dimensional waves that stand out on the picture. And add a few torn silver strips.

4 Using light brush strokes, brush the salt solution along the top of each blue wave. Once it dries it will leave frosted streaks.

Cosmic Gases

Oxygen is the main resource for this project. The project is a re-working of a favourite of pre-school children, but with a twist. To create these cosmic images use pale colours on a dark background; they will come together in a kaleidoscope of shapes. NASA images of supernovas and gaseous explosions are a good stimulus and can be found on the internet.

Resources

- Washing-up liquid
- Poster paints
- Black sugar paper
- Paint pots
- Plastic straws
- Water

Approach

1 Each container should be two-thirds full of water, squeeze in some washing-up liquid and some coloured paint as well as some white paint (so that the prints show up on the black paper). Stir with a straw.
2 Blow the liquid with a straw to create bubbles that reach the surface. Place the paper over the container horizontally to pick up the bubble prints, and then remove.

3 Repeat using the various available colours, each child should keep the same straw. A certain amount of guesswork will be needed when turning the paper over as they will not know where the print will appear, but this will create some nice surprises.
4 When the work is wet the picture will still be dark and it will appear to have not worked. However, once the picture is dry the colours will emerge.
   - Option: cut the paper into a circle for a bubble-shaped picture.

15

Flying Machines

This project explores historical flying machines through the medium of print. Look at images of hot air balloons, biplanes, modern aeroplanes and rockets and talk about how they manage to leave the ground and fly. Leonardo da Vinci's (1452–1519) drawings of flying machines (glider, parachute, and helicopter) are a fascinating prediction which lacked the technology to become a reality in Renaissance Italy.

Resources

- Large piece of paper or wallpaper
- Newspaper
- Poster paints
- Dishwashing sponges (unused)
- Cardboard
- PVA glue
- Glue sticks
- Printing inks, rollers and trays
- A4 cartridge paper
- Pallet

Approach

1 To make the printing block, use a piece of stiff cardboard (approx A4 size) as the base upon which to stick the pieces of cut sponge. Cut the sponge and assemble into a flying machine, and glue together using generous amounts of PVA glue. Put aside to dry.

2 Roll out the large piece of paper or wallpaper onto a surface the children can work around together. Draw a basic outline for the land/ cityscape that you wish to create.

3 Using the colours blue, orange, green, yellow and white in the pallet, apply the paint using the sponges for a stippled, textured affect. Allow to dry.

4 Draw and cut out shapes such as tower blocks, houses, boats and waves from the newspaper and stick onto the now dry panoramic picture using the glue.

5 Select strong colours to print with, roll generous amounts of ink onto the printing block. The nature of the sponge surface will mean that the ink appears not to adhere, but will in fact give you a nice print.

6 Start by printing on an A4 sheet of paper (on black if the chosen ink colour is pale). Then ink up again and print onto the painted panoramic. Continue to print on A4 sheets of paper using the variety of colours available. By this stage the colours will mix together as the children move from tray to tray.

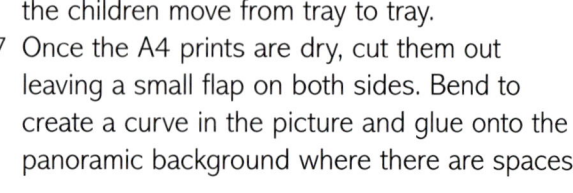

7 Once the A4 prints are dry, cut them out leaving a small flap on both sides. Bend to create a curve in the picture and glue onto the panoramic background where there are spaces.

Describing Evaporation

These handprints illustrate the gradual disappearance of an evaporating liquid. Taking inspiration from the work of artist Andy Warhol (1928–1987), some fluorescent colours have been used together with his technique of offsetting the second print on top of the first print to create an interesting effect.

Resources

- Poster or acrylic paints in light and dark colours
- A2 cartridge paper
- Plastic sheeting or tray

Approach

1 Choose a strong colour and squeeze a generous amount onto a waterproof surface or into a tray big enough for a hand to fit in to.

2 Encourage the children to spread the paint in the tray and press their hands into the paint, press their hands onto the paper and repeat from one side to the other, or in a circular formation until the paint fades away.

3 Ensure the children wash their hands.

4 Once the first set of hand prints are dry, squeeze the lighter, brighter colour in to another tray and press the same hand into the paint.

5 Invite the children to make another set of handprints on top of the first set following the same pattern, but this time place the hand slightly to the right of the first set, creating an 'offset' print. This makes a vibrant two-tone effect.

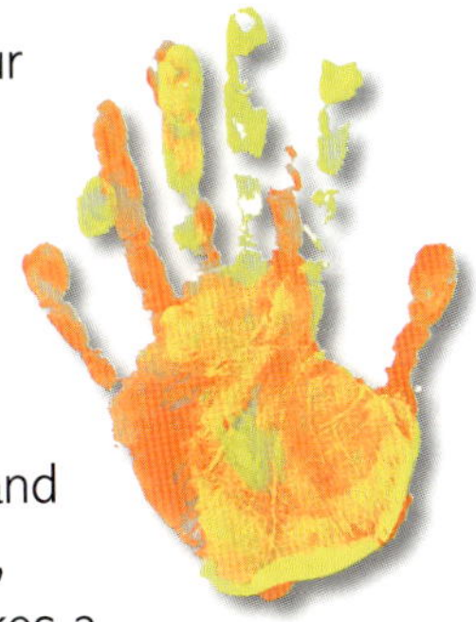

The Water Cycle

This project explores the water cycle by building a self-contained model showing the process of water evaporation from the seas, rivers and ponds. It also shows the water's journey up to a raincloud, and then falling as rain, eventually returning to the ocean.

Resources

- Stiff cardboard approximately 40cm x 25cm
- Blue, white and green acrylic or poster paint
- White flexible card or paper
- Blue paper
- Cotton wool
- Charcoal
- Pastels, crayons or coloured paper
- Glue sticks
- PVA glue
- Glue gun (teacher use only) or sticky tape

Approach

1 By applying blue, white and green paint, create a wavy texture over the entire surface of the cardboard base. Put it aside to dry. Once dry, the board may curl up and warp, but if bent gradually in the opposite direction it should flatten out again.

2 The cones represent the evaporation of seawater into the clouds. To make them, draw a circle approximately 40cm in diameter and cut the circle into quarters. Bend each quarter into a cone and staple the edges to secure it (teacher to do this).

3 Cut a thin strip of blue paper using a paper cutter and spread glue along the reverse side. Attach the end of the strip to the top of the cone and slowly wrap it around the cone in a spiral, sticking it down as you go.

4 For the cone that represents rainwater cut out small blue squares and stick them onto the cone.

5 To add the clouds, dip the tip of each cone into the PVA glue and put a piece of cotton wool on top of the cone(s) that represent evaporation. For the raincloud, take a small piece of crushed charcoal and rub onto your hands. Then gently darken some cotton wool between your fingers, and glue onto the top of the 'rain' cone.

6 Spread glue onto the rim of each cone and position, allowing to dry.

7 Lastly draw or collage a shining sun, cut it out and using a glue gun (teacher to do this) or sticky tape attach to the side of one of the cones that represents the process of evaporation.

Ammonites

Fossils are fascinating for children, and this project explores the process of fossilisation. It also demonstrates how plaster of Paris dissolves, and changes from a liquid to a solid in a matter of minutes. Gather images of ammonites and snails and look at the spiral shape that they both have. Real ammonite fossils can often be found in local museums with a natural history section.

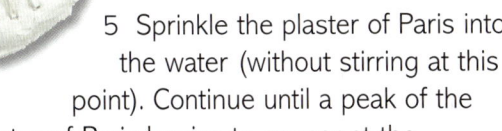

Resources

- Plasticine®
- Mark-making items, for example, lolly sticks, matchsticks, modelling tools
- Scrap card or a plastic tray
- Plaster of Paris
- Black poster or powder paint
- Water pots
- Water

Approach

1 Place the Plasticine® on the table and soften it by kneading. Once malleable, place onto the card or tray and form into a round.
2 Using fingers and thumbs create a recess inside the round so that it is a bowl shape with a completely flat base.
3 Using the collection of mark-making items poke the base to form a spiral pattern.
4 To mix the plaster, fill a water pot ¾ full with water, add a generous amount of black paint to it and stir.

5 Sprinkle the plaster of Paris into the water (without stirring at this point). Continue until a peak of the plaster of Paris begins to appear at the surface of the water. It will require five or six heaped tablespoons. Stir until smooth.
6 Without delay pour the dissolved plaster into the Plasticine® mould.
7 The piece will cure (set) within minutes. This is an interesting process for the children to observe. However, allow it to set firmly for at least a day before carefully peeling off the Plasticine®. To remove the Plasticine®, do so slowly and with care to minimise the loss of any detail. If it is done in a warm room or on a warm day it will come off very easily.
   • Option: lightly brush silver paint to the raised parts to add a shimmering effect.

Tropical Fish

Fish come in many shapes, sizes and colours. Look at several varieties of fish focusing on their unique patterns. This will be a starting point, but the scales of the fish can be modelled in an imaginative fashion. As well as demonstrating how plaster powder dissolves, the setting of plaster also shows an irreversible change. This project can be used to make work to support any other curriculum area or festival, for example, a Tudor rose, Egyptian tablets or Christmas decorations.

Resources

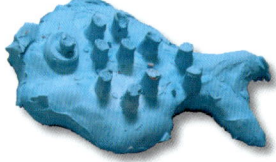

- Plasticine®
- Mark-making items, for example, buttons, corks, sticks, straws
- Scrap card or a plastic tray
- Plaster of Paris
- Powder or poster paint in a variety of colours
- Water pot
- Water
- 10cm long strong fabric strips or shoelaces

Approach

1 Place the Plasticine® on the table and soften it by kneading. Once malleable, place onto the card or tray and form into a simple solid fish shape around 5cm in height.
2 Using fingers and thumbs create a recess inside the fish so that it becomes a vessel. Using the collection of mark-making materials poke the base and sides to create an interesting surface pattern.
3 Fill a water pot ¾ full with water and start

to sprinkle the plaster of Paris into the water (without stirring at this point). Continue until a peak of the plaster of Paris begins to appear at the surface of the water. It will require five or six heaped tablespoons.
4 Add a tablespoon of powder paint, and slowly stir the mixture using a stick or spoon until smooth.
5 Without delay pour the dissolved plaster into the Plasticine® mould.
6 The piece will cure (set) within minutes. This is an interesting process for the children to observe. However, allow it to set firmly for at least a day before carefully peeling off the Plasticine®. To remove the Plasticine®, do so slowly and with care to minimise the loss of any detail. If it is done in a warm room or on a warm day it will come off very easily.

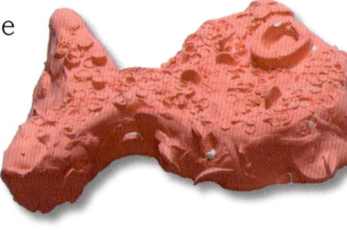

7 To display, arrange as a shoal of fish on a flat surface.

Salt Dough Street

This project demonstrates the irreversible change that occurs when applying heat to dough. By baking the sculptures for different lengths of time, the overall effect gives the pieces different tones, from light to dark.

Resources

- Modelling tools and/ or plastic knives
- Salt dough mixture
- Plain flour
- Containers
- Greaseproof paper
- Water
- Salt

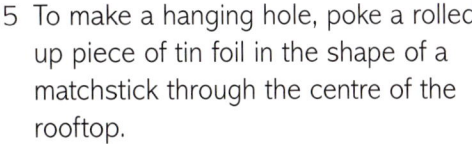

Approach

1 **Salt dough instructions:** place one cup of flour, half a cup of salt, and half a cup of water in a container. Mix together until smooth, adding small amounts of water if it is too dry.

2 Once the dough is ready, form a springy ball. If it is not too sticky or too dry, it is ready to work with. Place on the greaseproof paper to begin working, this will prevent the work sticking to the table.

3 Using an adult palm-sized ball of dough, flatten it to 2–3cm in thickness and begin to shape it into a house. The tools and knives are useful for cutting straight edges for the roof.

4 Use the tools to make the details, such as the roof tiles and windows. Do not press too deeply when drawing the details as this can cause the dough to snap once baked.

5 To make a hanging hole, poke a rolled up piece of tin foil in the shape of a matchstick through the centre of the rooftop.

6 **Baking:** bake the pieces on a baking tray at a low oven temperature (100–150°C) until hard, approximately 1 hour.

7 The length of time the pieces are baked for should vary, pieces taken out first will be the palest, those that follow will be darker, and those taken out last of all will be very dark. To ensure variety, take pieces out at different times. Remove the foil matchsticks once baked.

8 Display the houses in the form of a street, varying the pieces to show the different tones achieved by the process.

Kaleidoscopic Self-Portraits

These self-portraits are a quick, clean and fun way to create an interesting 'fragmented' picture. For young children, putting together the pieces is a jigsaw-like challenge. The squares can be made even smaller for older children for a bigger challenge and to achieve some interesting visual results.

The method is based on British artist David Hockney's (1937–) photo-collages. A good start is to show the class images to get their responses. The fragmented images often look as though there are many more people than there are. They contain a lot of movement and show an image from many angles. Picasso's (1881–1973) *Portrait of Dora Maar Seated* (1937) tries to show many aspects of the face in one picture.

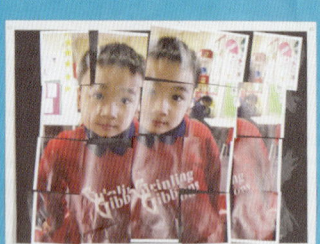

Resources

- Colour (or black and white) photographs of each child (x2 copies)
- Black or white paper
- Glue sticks

Approach

1 Cut each portrait into 12 equal squares and give to each child.

2 Allow each child to reassemble their portrait like a jigsaw. Before sticking the squares down, hand out the duplicate set of squares and encourage the children to include some of them in order to alter the appearance of their portrait.

3 Taking inspiration from David Hockney's photo-collaged portraits, you can add an extra set of eyes or another mouth. Glue into position onto the paper.

Going Places

This project is quick, simple and full of humour. Children can use their imaginations to change how they look, where they are and explore the possibilities of who they want to be. One artist who recorded himself ageing was Dutch painter and etcher Rembrandt (1606–1669). Some of his self-portraits are dramatic, others are comic and humorous and he uses different kinds of clothing to change his appearance. His self-portraits also reflect a different mood at each stage of his life. As a young man he is fresh-faced and ambitious, and as an older man he appears thoughtful and weary. These ideas can be explored as a start to the project, with the children using mirrors to look at expressions and feelings.

Resources

- A4 black and white photograph of each child
- Oil pastels

Approach

1. Use the oil pastels to add colour to the surrounds of the face and the clothes.
2. Encourage the children to verbalise what they would like to be when they grow up, and what they would wear for different professions. Use imagery of doctors, police and so on, if necessary.
3. The same technique can be used to explore ideas of place. The children can transform the background to show themselves in a different place like the beach or in space, for example.

Your Heart

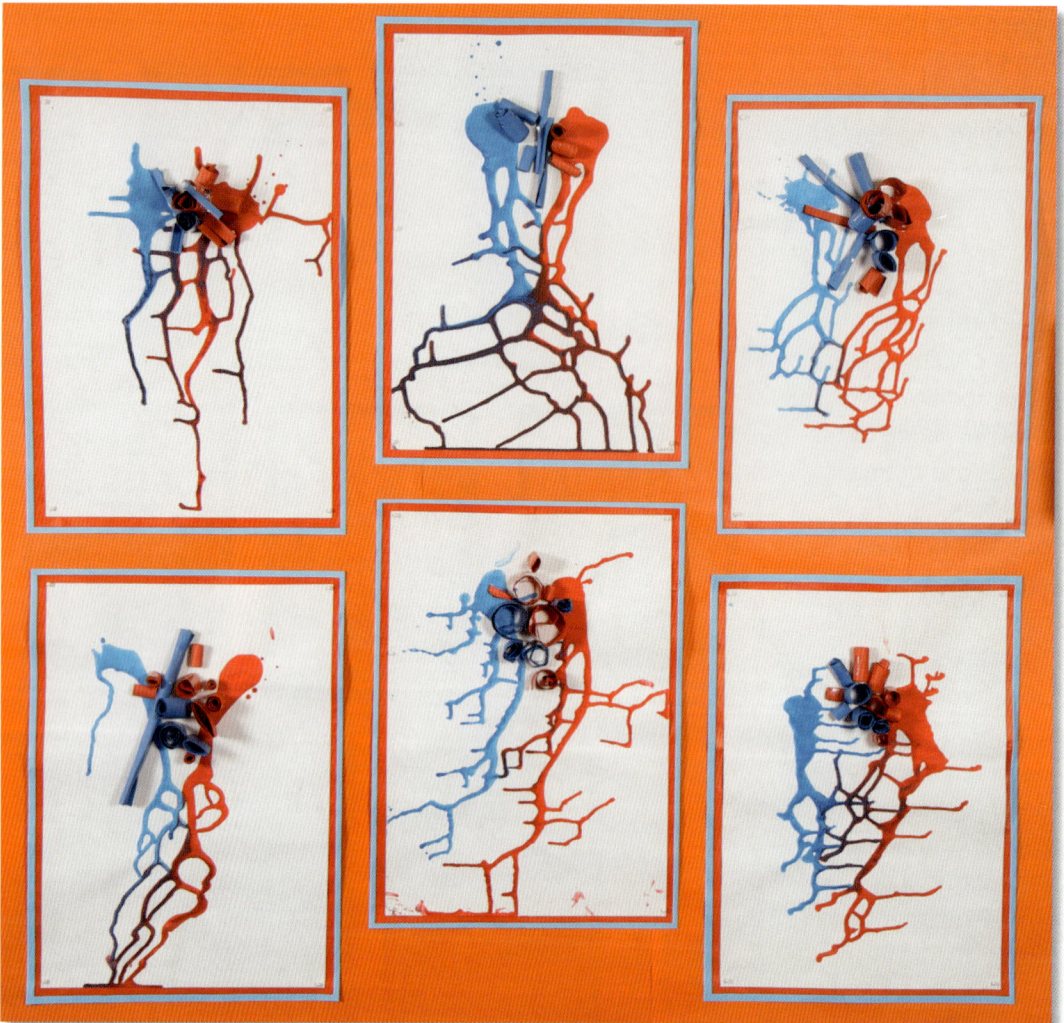

This simple collage using flowing paint, is an illustration of how blood circulates round the body. It is an interpretation (not an exact representation) of the how veins and arteries are structured within our bodies, carrying an oxygenated and deoxygenated blood supply to and from our vital organs. Begin by showing the children diagrams which illustrate how the heart pumps blood around the body, and the difference between veins and arteries.

Resources

- Red and blue acrylic or poster paint
- Red and blue paper
- A3 cartridge paper
- Large brush
- Pallet
- Glue sticks
- PVA glue
- Scissors

Approach

1 Dip the brush in the blue paint and let it drip to the left of centre on the paper, preferably without making contact with the paper.
2 Wash the brush and repeat using the red paint dripped slightly to the right. Then begin to tip the paper slowly in different directions to create a web of veins across the sheet. Allow the colours to merge a little as they meet and criss-cross. Allow to dry.
3 To make the heart with arteries and ventricles, cut strips of red and blue paper in varying widths. Roll up and secure the scroll using the glue stick. The coils can be gently bent into various shapes as well as being circular.
4 Use the PVA glue to fix each scroll in the centre of the paper to make the heart. Use large coils to make the ventricles, and then use longer, thinner scrolls on their sides to make the major arteries and veins.

Pulse Pictures

For inspiration look at images by the Italian artist Guiseppe Arcimboldo (1527–1593) who painted clever illusionistic portraits of fruit and vegetables, and Paul Cézanne's (1839–1906) still life paintings that are bright and colourful. Taking the idea of 'you are what you eat' a little further, this project uses three very different media – photography, oil pastels and collage – to great effect. To begin, ask the children what their favourite sport or fitness activity is.

Resources

- A3-sized pieces of cardboard
- A3 prints of each child
- Digital camera
- Glue sticks
- Colour oil pastels or crayons
- A variety of differently coloured pulses, beans and pasta shapes

Approach

1 Photograph each child in their keep-fit pose and print out on A3 paper. Using the glue stick rub an even and generous amount of glue onto the cardboard, position the picture and flatten down.

2 Ask the children to imagine where they are whilst keeping fit (for example, winning a sports competition or out in the countryside) and transform the surrounds of the picture using colour pastels.

3 Dribble generous amounts of PVA glue and use the beans, pulses and pasta to enhance the figure in the picture. Only use one type of material for one area, for example, outlining the edge of the figure, or filling in one garment of clothing. If too many materials are used the picture can look confused. Allow it to dry flat.

4 For display purposes and to minimise the materials falling off, use a large paint brush and apply PVA glue over those areas. It will dry clear and strengthen the piece.

Hard Shell Invertebrates

This project illustrates the properties of invertebrates with a hard outer shell and a soft internal structure, by using very hard and very soft materials together. Begin by discussing the differences between our bodies and those of hard shell invertebrates, and talk about how and why a hard shell might be useful for the habitats they live in.

Resources

- Glue gun (optional) – adult-use only
- Balloons
- Mod-Roc
- Water and containers
- Wadding
- PVA glue

Tortoise
- Natural coloured tights
- Googly eyes
- Green, brown, yellow and black tissue paper
- Brown or gold paper
- String or elastic bands

Ladybird
- Black tights
- Black card for legs
- Red, yellow and black tissue paper

Crab
- Natural tights
- Googly eyes
- Pipe cleaners
- Yellow, orange, pink and gold card or similar

Approach

Shell:

1 Inflate a balloon to approximately 15cm in diameter. Put to one side. Cut a pile of Mod-Roc strips each approximately 6cm in width.

2 Show the children how to use the Mod-Roc, dipping it in the water once, allowing the excess water to run off. Knead to soften it and place at the bottom of the balloon. Build up the Mod-Roc from the bottom to halfway up the balloon with at least two or three layers for strength. Covering the bottom of the balloon will produce an even circle, as opposed to the pear drop shape of the side. As the balloon will be tricky to hold, a small bucket or deep container to place it into whilst working will help stabilize it.

3 Allow at least one day for the Mod-Roc to set hard, remove the balloon, trim the edges with large scissors and shake off the excess particles into the bin.

Colour:

1 Tissue paper is used to add colour to the bowl in layers. Encourage the children to look at the pattern markings of each creature and recreate them using layers of colour and pieces cut to shape. Brush the surface area of the bowl with the PVA liquid. Brush on an additional layer of the PVA mixture to add shine to the surface when dry.

2 Use the same method as above but suggest pattern-building in stripes or patchwork, choosing a specific palette range, for example, shades of greens or brown for the tortoise, pink, peach and yellow for the crab.

3 To attach the legs and claws, use PVA glue and place a little sticky tape or masking tape to hold them in place and allow to dry upside down. Alternatively, glue into position using a glue gun (teacher to do this). Stick on googly eyes to finish.

Soft internal sections:
Tortoise

Cut one leg from the tights and stuff with the wadding to form the head and four legs. Tie the toe-end tip of the stocking around the neck of the tortoise in a knot. Tie with string or elastic bands to form the legs. Tie the open end around one of the back legs. Spread glue in the underside of the tortoise shell, place the soft body inside and allow to dry upside down.

Ladybird and crab

Using the top part of the tights, cut off the legs and stretch the nylon widthways. Stuff the nylon with large pieces of wadding. Tie the ends in a double knot, or staple the ends closed. Glue in place as above.

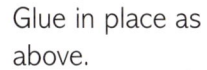

Soft Invertebrates

This project uses soft materials for soft invertebrates. Start by looking at images of jellyfish and starfish. Discuss what the structure of the creatures appears to be and how they move around. The creatures are delicate and the children should try to capture their ethereal quality.

Resources

Starfish
- Tights
- Acrylic paint in a variety of colours
- Pipe-cleaners, string, elastic bands
- Scraps of fabric or felt
- Wadding
- Large buttons

Jellyfish
- Crêpe or tissue paper
- Metallic paper
- Pipe-cleaners
- Fluorescent netting or a plastic shower cap

Approach

Starfish

1 Cut the full leg length of the tights into five equal pieces, securing one end of each piece with a knot.
2 Squeeze the acrylic paint of choice into the container with a little water to thin out and stir with a paint brush. Dip the stocking into the paint mixture and stir around, using your hands to get an even coating of pigment. Remove from the paint mixture and flatten out, then hang on a coat hanger to dry.
3 Once dry, put both hands into the opening and gently stretch width-ways to widen. Stuff with wadding to make the leg shape. Tie a knot to secure the end once full. Repeat this five times so you have enough legs for the starfish.

Jellyfish

1 Using four or five pipe-cleaners, twist them together to create a circle. Use the remaining pipe-cleaners to make a dome shape.
2 Tear lengths of crêpe paper and cut lengths of metallic paper into thin 2cm wide strips. Bend the metallic strips to create a zigzag affect. Bend the tips of the paper strips over the edge of the pipe-cleaner and secure with glue, letting them hang down.
3 For the outer part of the head, cut a piece of netting approximately 40cm x 40cm and secure it by stapling it around the edges of the dome shape (teacher to do this). For the bright internal section, make a circle from pipe-cleaners and cut a piece of fluorescent netting so that it is like a hat, fold it under the circle and staple all the way round to secure it in place (teacher to do this). For the bright part of the jellyfish scrunch the smaller piece of netting and place inside the 'hat'.

Faster! Slower! Drip Paintings

This project explores what happens when force is applied quickly and slowly to the moving paint across paper. The changes in direction create some interesting and unpredictable lines and shapes. New colours emerge as the streams of paint travel across the moving page. The eyes are a fun way to show the direction of the drips, as they look up and look down and to the sides.

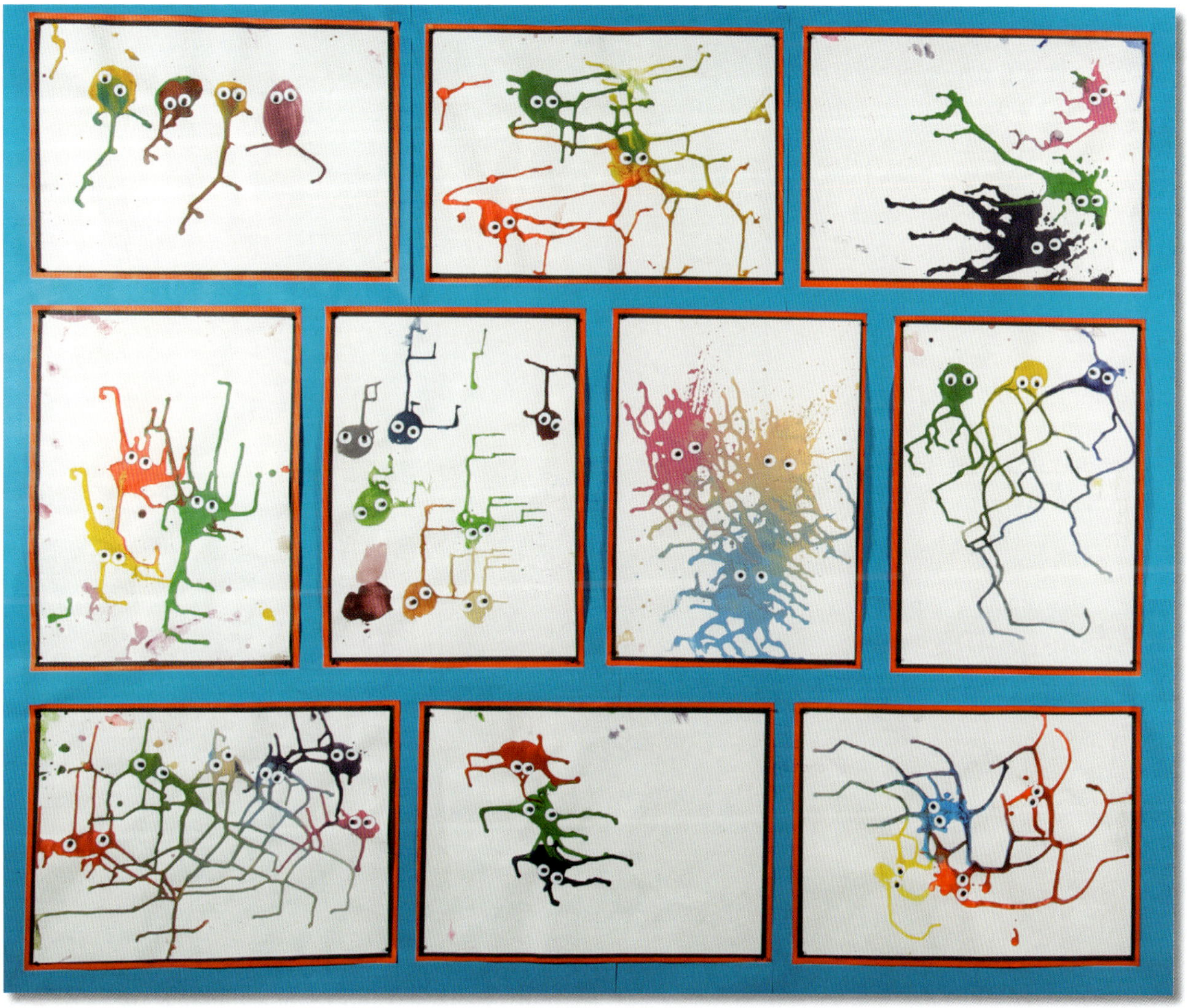

Resources

- A3 cartridge paper
- Acrylic or poster paint in bright colours
- A large brush
- Pallet

- White and black paper, circular stickers or googly eyes

Approach

1 Dip the brush in the first colour and allow it to drip onto the paper, preferably without the brush making contact with the paper.

2 Wash the brush and repeat the process finding a new space each time. Then begin to tip the paper slowly and quickly to create interesting marks across the paper.

3 Once dry add the eyes. The easiest way to make these is to use white and black stickers, one on top of the other, or to use googly eyes. Alternatively, cut white circles for the eyes, and punch holes in black paper, collecting the punched pieces to use as the pupil of the eye.

Push and Pull

This project explores the forces of push and pull. When making the piece the children will discover other forces such as rolling, squeezing and squashing pieces of tin foil, bending and twisting foil and pipe-cleaners and bending and threading plastic tags. Young children can enjoy a show and tell of the forces they have used to create this work. For an artistic focus, look at the work of Cornelia Parker (b. 1956) who flattens metallic objects like silverware and musical instruments and suspends them with invisible string. They are suspended using the force of gravity and show the force used to flatten each object.

Resources

- Shoebox lids
- Acrylic paint
- Paper craft straws
  (5 per person)
- Glitter (optional)
- PVA glue
- Tin foil, pipe-
  cleaners, plastic
  tags, and so on
- Coloured paper

Approach

1 Paint the inside and outside of the shoebox lid in one colour of choice.
   - Option: whilst the inside is still wet sprinkle a pinch of glitter onto the surface.
2 Once dry, apply a generous amount of PVA glue to both vertical edges of the lid, and lay the five straws in your choice of colours across the lid and allow them to dry. Once dry, cut off the overlapping straw at the sides.

3 Create shapes with the foil, such as circles flattened and bent around a straw, or twist the foil in a spiral around another straw. Create pieces that can slide up and down. Add pipe-cleaners and plastic in a similar way, other suitable materials will work too.
4 The finished piece may resemble an abacus with parts that can be moved left to right along the straws.

Earthquake City

This project is a two-in-one science activity as it uses springs and demonstrates how an earthquake works. As a starting point look at images of cities damaged by an earthquake and ask the group how they think an earthquake does this.

Resources

Ground structure
- Four 15cm lengths of fine modelling or gardening wire
- Two pieces of cardboard 30cm x 30cm
- Map
- Colouring pencils
- Glue sticks
- Black or grey tissue paper or paint
- Bright paper

City
- Flexible card
- Sponge pieces
- Acrylic or poster paint in dark colours
- PVA glue
- Digital camera (optional)

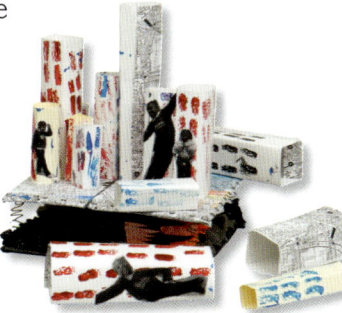

Approach

Ground structure
1 Photocopy a map of a town or city enlarging it to fit a sheet of A4 paper. Colour in areas such as the streets, rivers and parks.
2 Completely wrap one of the pieces of cardboard in the map, gluing it on the underside to hold it in place.
3 The second piece of cardboard will be the cracked earth. Wrap with black paper, tissue paper or black paint, gluing on the underside. Then using bright orange paper, cut a zigzag shape and glue towards one edge.
4 The four springs are the separating mechanism. Wind each piece of wire around a pencil to make the springs. Using the end of the spring, poke it into each corner of the cracked earth and the street section.

The City
1 To make the tower blocks cut a piece of flexible card to your choice of height, the card must be wide enough to fold four times to create a cuboid – leave a small flap on one edge to glue the oblong into shape.
2 Make as many buildings as wanted, all at varying heights.
3 Wrap some of the buildings with pieces of the map. To make the windows use a small piece of sponge and dip into the paint and dab onto each side of the building.
4 Select some buildings that will remain stable during the earthquake, apply PVA glue to the bottom edge and allow them to dry. Position the remaining buildings in between, give it a wobble and watch the earthquake in action!

31

Rough and Smooth – Collage

Create a piece of work that is as interesting to touch as it is to look at, by using rough and smooth materials. Autumn leaves lend themselves well to this textural piece, but a seascape with crashing waves in velvet and silk against grey rocks made of crinkly paper would work equally well.

Resources

- Black sugar paper
- PVA glue
- Wood sticks, paper straws, thin paper strips
- Hessian
- Tissue paper
- Newspaper
- Silk
- Metallic paper
- Tracing paper
- Sandpaper
- Velvet
- Woollen fabric

Approach

1 If available, bring in real autumn leaves, focusing on the rich colour, crinkled texture and curled-up shapes. If real leaves are not available, find a variety of pictures for the children to copy.

2 Draw around a leaf on flexible card and cut it out to create a stencil.

3 Using the sticks, straws or strips of paper create a tree with bare branches.

4 Cut squares approximately 6cm x 6cm from the variety of materials and papers, and using the stencil draw around and cut out each paper or fabric leaf. Scrunch up some of the paper leaves to create a crinkled effect.

5 Using PVA glue begin to build up the collage overlapping the leaves, putting glue only on part of the leaf, allowing it to curl away from the paper.

Rough and Smooth – Containers

This project offers a great deal of scope for exploring different textures and surfaces. The suggested list of items to imprint on the clay is a starting point; many other suitable items can be used. The assembled tiles create a finished 3D container, however, the separate tiles are also very attractive.

Resources

- Air-drying clay
- Modelling tools
- Water pots
- Canvas or J-cloths™
- Embossed wallpaper
- String
- Sticks, twigs, matchsticks
- A wooden shed
- Playground surface
- Playground wall
- Tree bark
- Metallic paints
- Poster paints in rich colours
- PVA glue

Approach

Ground structure

1 Roll out the air-drying clay into slabs and cut it into five squares 15cm x 15cm, or smaller.
2 Place them in turn on a surface such as a wooden board, or plastic sheeting with a piece of canvas or J-cloth™. This will stop the clay sticking to the table.
3 For two of the slabs choose one of the textured 'indoor' materials and press it into the clay.
4 For the other two slabs choose a surface from the 'outdoor' materials and press the clay onto it.

5 To assemble the five slabs into a cube-shaped container, use a modelling tool or fine stick to scratch all four edges of one slab, and three edges of the vertical slabs that will be attached to one another.
6 If by this stage the clay has become dry, use a little water to moisten the edges. Pinch the joins together, all the while taking care not to rub off the impressions made earlier.
7 For added strength make a thin sausage of clay and gently press it into the joins inside the pot. Allow it to dry completely.
8 Use two coats of metallic paint to decorate the outside of the container. Then use two coats of poster paint for the inside. Once dry, varnish just the inside with two coats of slightly watered-down PVA glue. Allow to dry fully.

Making a Splash

This is a very short and surprisingly clean activity that uses just paint and gravity. One inspirational artist is Jackson Pollock (1912–1956) who during the 1950s enjoyed dripping and splashing paint directly onto a canvas on the ground. Footage of him working and talking can be seen on the internet. The same activity within the classroom is just scaled down using a piece of paper and standing up whilst working.

Resources

- A3 black or white paper
- Poster paints in 600ml tubes

Approach

1 The children should work standing up, holding the poster paint upside down and allowing it to drip. By squeezing the tube lightly and moving quickly guide the paint across the paper to make lines, circles and free-form shapes.

2 Change colours frequently, building up an interesting sheet of colour.

3 Allow to dry flat, taking care when moving to a drying rack so that the colours do not run. The pictures will take a few days to dry completely and need to be handled with care so that the paint does not crack and fall off.

Powder Power

These pictures are created using weight and the impact of powder paint dropped from above. The powder is contained so there is limited dust, and it is a short activity that demonstrates weight and gravity in a colourful way.

Resources

- White paper (A3 or A2 size)
- Powder paints in a variety of colours
- New pop socks
- Water sprayer
- Laminator and A3 or A2 pouches

Approach

1 In preparation, spoon a heaped tablespoon of powder paint into a pop sock. Repeat for the other colours. Each pop sock should contain a different coloured paint.
2 Spray the sheet of black paper with water using the spray gun.
3 Holding the pop sock at the top, stand up and drop the sock onto the wet paper. The paint will disperse and make a powder puff, and the water will hold the powder in place.
4 Change colours to create an interesting image. Then spray water lightly on top of the finished piece and allow it to dry.
5 Once dry, laminate the piece to prevent smudging, remove crinkles and intensify the colours.

Magic Magnets

This project offers three possibilities based on the same principle. By moving one magnet on the underside of the board the magnet on top of the board appears to travel around by itself. You can create your own setting and make different moving parts. These three ideas are a starting point. And the simplest of materials can be used to make it colourful.

Resources

- Two small magnets
- Stiff board, for example, mount board approximately A4 size
- Coloured pencils and pens
- Coloured paint
- Flexible card
- Metallic paper
- Graph paper
- PVA glue or glue gun (teacher use only)

Approach

Map and bus

1 Enlarge a section of a map and glue it to the board, trimming off the edges.
2 Colour in the main streets.
3 For the bus, make a cuboid using flexible card, and paint the bus in a chosen colour.
4 On small square pieces of paper draw the passengers, the driver and the bus number, and so on.

Rocket in space

1 Draw and paint a space scene with planets and stars.

2 Using the metallic paper, draw around a circular object and cut out the shape to make the planets.
3 To create the rocket make an oblong with flexible card, add a cone tip with metallic paper and attach with PVA glue, leave it to dry.

Adventure maze

1 Draw the route of the maze on graph paper. The route should be 1cm in thickness.
2 Paint the route and once dry draw a neat edge around it.
3 Draw pictures in the spaces to create the adventure land.

To make the moving component

1 To attach the magnet turn the bus/rocket/character over and put a generous amount of glue on the underside towards the front. Place the small magnet on the glue spot and allow it to dry for several hours.
2 Once dry, turn over and place the bus/rocket/character onto the board and place the second magnet underneath until they are 'attached'. You can then move the object around.

One Giant Step

The iconic images from the Apollo Moon landing are a fantastic starting point for this journey to the Moon, and you can find the images on the internet. Discussion points could include the effects of the lack of gravity, the Moon's atmosphere, identifying things that look different to photos that are taken on Earth. Also look at the casts by artist Rachel Whiteread (1963–) which are ghost-like impressions of an object that has been removed. This project uses plaster casts as proof that someone has walked on the moon by making a mould and taking a cast.

Resources

- Clay
- Containers, for example, foil trays, margarine tubs
- Modelling tools
- Greaseproof or tracing paper
- Five cups of plaster of Paris
- One litre of water
- Pourer or jug
- Bucket
- Hessian strips or scrap fabric strips

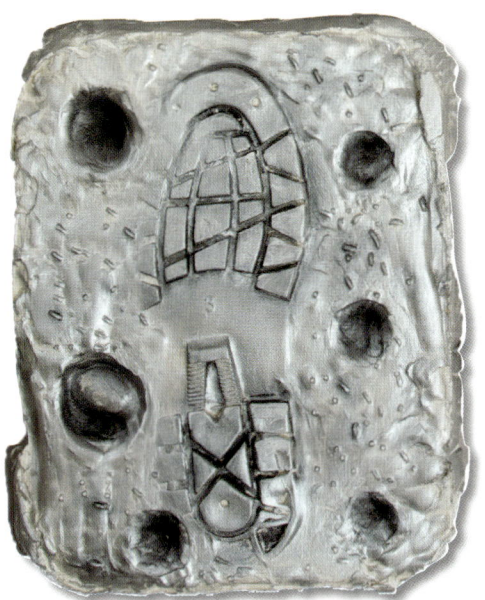

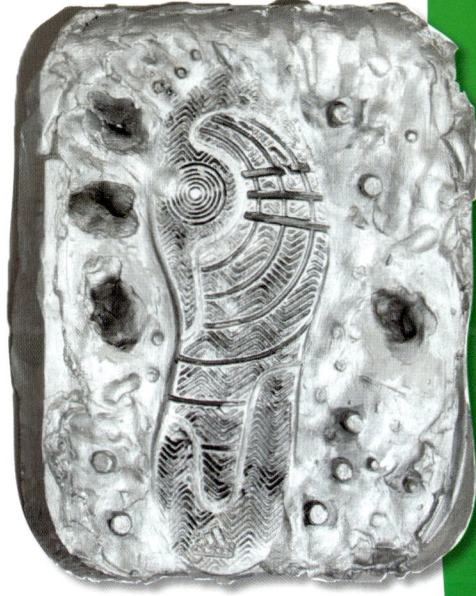

Approach

1 Start with a container that is big enough to take a child's shoe or hand, leaving space around the object. Line the bottom with paper to prevent the clay sticking to the tray.

2 Take a lump of clay and roll it out to the dimensions of the tray and about 5mm thick. Place inside and ensure it fills the corners.

3 For a foot impression place the heel into the clay first, then the sole. Finally press the toes down at the tip of the clay. For an impression of a hand or fingers, place the hand centrally in the clay and press down, then model the clay up around the fingers almost covering the back of the hand, then remove. This will create a bolder impression once cast in plaster.

4 Leave the imprint as it is, but work around it to create a pitted surface by poking the modelling tool in the clay. For craters add small balls of clay.

5 For the casting stage, slowly sprinkle each cup of plaster in the centre of the container one by one. Stir with a spoon removing the lumps. Pour the plaster without delay into the tray filling the depth (minimum 3cm), then gently submerge the hessian or fabric strips as this will strengthen the cast. Tap the sides of the tray allowing the air bubbles to rise. Only remove from the tray after one day.

6 Remove the plaster block by turning it over with care and allowing the slab to dislodge. Peel back the greaseproof paper, and pick out the wet clay and throw it away. Rinse the slab and brush away any clay residue.

7 Paint using black paint for the craters and recesses, then blend with silver keeping the brush as dry as possible. Brush on more silver for a shimmering piece of moon rock.

Lunar Drawings

These lunar drawings create some of the wonder of space and a sense of infinity, but on a small scale. On the internet you can find images of astronauts floating in space and repairing satellites, and different modes of space travel over the years. These make perfect silhouettes to draw around, and are great images in themselves.

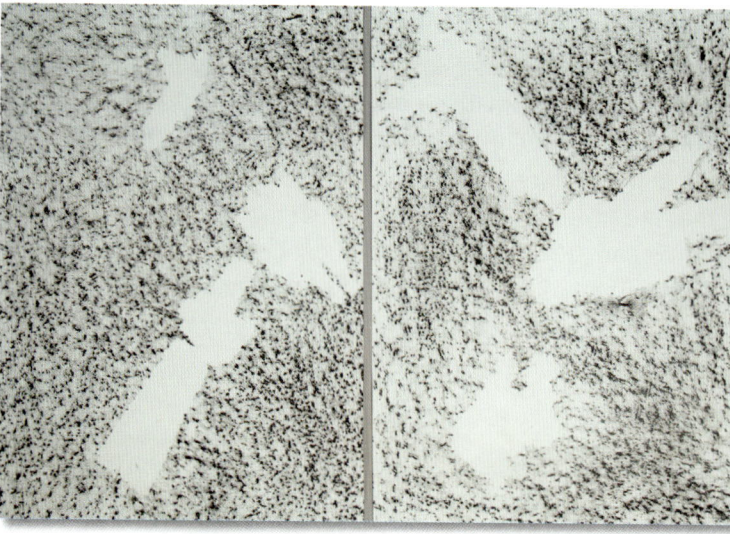

Resources

- A3 white paper
- Black crayons or graphite sticks
- Photocopied images of satellites, astronauts, space rockets/shuttles
- Scissors
- Masking tape or Blu-Tack®
- Clear plastic pots
- Silver metallic paper
- Sticky tape
- Stapler (teacher use only)
- Silver paint

Approach

1 Carefully cut out three images from the selection of space pictures. Position the paper horizontally and stick the images on with the masking tape or Blu-Tack®.

2 Place the paper on a rough surface (outside in the playground, for example) or on the table and using the side of the crayon or graphite stick rub it on the paper, filling in the white space. Rub quite heavily at the edge of each image as this will give the silhouette definition. When finished, remove each image to reveal the silhouettes.

3 Cut a piece of silver paper the depth of the plastic pot and line the pot with it. Use sticky tape to hold it in position.

4 Bend the paper into a cone shape, overlapping the two bottom corners. Drop the plastic pot down into the cone to check that it fits in the hole. The diameter should be wide enough so that the pot drops nicely through the hole, but that the rim of the pot sits at the edge of the cone opening. Now staple the pot into position (teacher to do this). Then put sticky tape on the outside of the join to secure it.

5 Hold the cone up to the daylight and turn it to see the light reflect off the surfaces.

Renaissance Inventions

The exciting work of Renaissance artist Leonardo da Vinci (1452–1519) provides the basis for this project. He lived in Florence, Italy, and was a painter, sculptor, architect and inventor fascinated by the world around him. His drawings show his ideas for flying machines, a parachute, diving suit, a tank and a war machine. These inventions could not be realised due to the insufficient technology of the period. Look at da Vinci's sketches of people, portraits, and his ideas sheets. This project uses wood as that was the only suitable construction material available at the time.

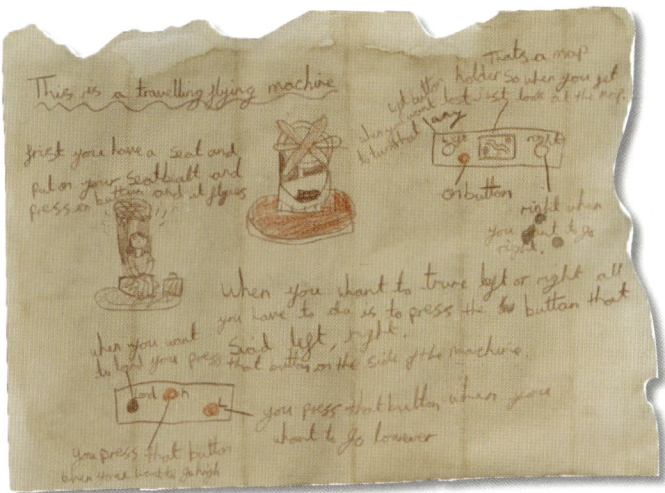

Resources

- Lolly sticks and/or brown art straws
- Small recyclable items, for example, plastic pots, boxes, corks, paper plates, paper
- PVA glue
- A4 cartridge paper
- Brown or lead pencil
- Instant coffee granules

Approach

1 Dilute some instant coffee granules in cold water, leave to one side. Tear the edges off two sides of the cartridge paper and brush the cold coffee across the paper, put aside to dry.

2 Take one or two of the recyclable items and begin to assemble them in an inventive way with the PVA glue. Cover the recyclables with the lolly sticks. If any slip off they can be temporarily taped together with masking tape whilst drying.

3 When the paper is dry use it as an ideas sheet, sketching diagrams as to how the invention works. Instructions and explanations can be added to the sheet.

4 The drawing and the model can be worked on at the same time, one supporting the other as ideas develop.

Mission to Mars

Space travel began in the 1950s and spawned many science fiction films and comic strips, especially in the USA. This project is a nod to that time, making a shiny rocket fit for the outer reaches of the universe.

Resources

- 1.5 litre or 75ml plastic water bottles
- Metallic paper
- Glue sticks
- Sticky tape
- Paper cutter
- Stapler (teacher use only)
- Glue gun (optional) (teacher use only)
- Fishing nylon (optional)
- Flexible card

Approach

1 Remove the label from the bottle and wash with a little warm, soapy water.
2 Using a paper cutter, cut strips of the metallic paper approximately 1cm wide, and a few that are wider.
3 Begin to stick the strips around the bottle using the glue stick, leaving the occasional space between each strip if desired. Make sure all the joins are on the same side. The glue goes on each end of the strip, which is trimmed to fit around the bottle.
4 The glue will only hold the strips temporarily.

Once all the strips are on the bottle put a line of sticky tape over the joins to secure the paper to the bottle. This is the back of the rocket.
5 For the rocket booster use a piece of flexible card cut into a circle, cut a slit to the centre and bend into a cone shape, staple into position (teacher to do this). Press the tip of the cone inwards to make a dent; this will be a more stable platform for the rocket to be glued onto. Use a glue gun (teacher to do this) or sticky tape to glue the two pieces together.
6 For the rocket 'nose' cut a circle from the metallic paper with a larger diameter than that of the booster base. Cut a slit to the centre and bend into a cone shape. Apply the glue gun (teacher to do this) to the tip of the bottle and place the cone on top. Let the glue cool.
7 To make the wings, cut diagonally across the width of a rectangular piece of metallic paper. Fold back the longer edge of each wing to create a flap and glue or tape this flap to the bottle.

Nocturnal Animals

This project involves creating a mini-forest for nocturnal animals to inhabit. Begin by looking at images of night-time animals including hedgehogs, foxes, owls and bats. A torch through the small window gives a glimpse into the night-time activities of these creatures and can be added to and altered. As a project for young children draw a fox template for them to work from.

Resources

- Shoebox without lid
- Black acrylic paint
- Dark blue acrylic paint
- Selection of colouring and collage materials, for example, metallic and tissue papers, cotton wool, Plasticine®, glitter, matchsticks, foil, small buttons, colouring pencils.
- Flexible card
- Glue sticks

Approach

1 Paint the interior of the box dark blue and the exterior of the box black, and put aside to dry.

2 Use the collage materials to make and decorate the fox, hedgehog and Moon.

3 Draw around a simple fox template, colour the fox in using orange pencil, crayon, or paint.

Add cotton wool for a bushy tail. Bend the flap at the bottom and add some glue to it.

Stick the fox into position in the box.

4 Roll a small piece of Plasticine® into a ball. Cover several headless matchsticks with tin foil. Then poke them into the Plasticine® ball, making sure they are evenly spaced, to create a silver hedgehog. Pinch a nose using a finger and thumb, and use two buttons for eyes.

5 Partially cover a button with tin foil and glue to the back of the box with PVA glue.

6 Finally, make a 'curtain' cover using black sugar paper. Cut out stars to look through. Drape the 'curtain' across the front of the box, make a fold at the top and glue it to secure it in place. Put the box on a table and shine the torch through the cut-out star shapes so the children can see what they can find in the forest!

Lighting Up the Night Sky

Fireworks fill us with awe and reflect a time of celebration. The coming of the New Year, Bonfire Night, Diwali and the Chinese New Year are just some of the festivities that occur in many towns and cities all around the world where fireworks are used. This project uses light and reflective materials to capture the moment when a town or cityscape is illuminated by the magic of fireworks.

Resources

- Shoebox
- Silver and/or gold paper
- Fluorescent Post-it Notes® or paper
- Battery-operated fairy lights
- Lolly sticks or card strips
- Dark grey paint
- Black sugar paper
- PVA glue

Approach

1 Paint the inside and outside of the shoebox grey and allow it to dry.

2 Draw and cut out a variety of houses and rooftops that are flat, triangular and so on, each should be a third of the height of the box. Stick each one onto a piece of gold or silver paper, trim leaving a 5mm edge, fold a flap at the bottom and stick into the box.

3 Start by making a base for each firework. Glue three lolly sticks together in the centre to create a star shape and allow them to dry.

4 For the firework explosions, tightly roll a Post-it Notes® at the corner to make a cone (a tube shape will also work). Glue one to each arm of the lolly stick base.

5 Add shine to each firework by making cones from the metallic paper. Put glue on the tip of each one and gently place in the centre of the lolly stick base.

6 To finish glue the fireworks to the inside of the box. Leave to dry.

7 To illuminate the piece, poke small holes near the centre of each starburst and slot in the fairy lights one by one, leaving the battery section at the back of the piece. Turn the lights on to see the reflective materials shimmer in the night sky.

Earth as Pigment

This project uses soil to create a painting medium. It also explores the process of germination, roots and shoots and looks at all the things plants need to grow. Start by looking at photos of growing plants that show the network of roots below the soil and the budding leaves. Explain that new leaves are a lighter shade of green and that the spread of roots underground spreads out and downwards from a central point.

Resources

- A cup of soil
- PVA glue
- Heavyweight cartridge paper
- Green art straws, paper or fabric
- Blue watercolour paint
- String
- Water

Approach

1 To prepare, dry the soil and remove any stones or roots. Push the soil through a large sieve and then through a finer sieve or strainer to create a fine powder. Put to one side.

2 The paper should be portrait in orientation. Draw a horizon line. Wet the paper above the horizon line and briskly drag the blue paint across the damp paper and allow the colour to spread out naturally. Apply a little more if necessary.

3 Add water to the soil and stir into a paste. Add a little PVA glue and stir in.

4 Using a large paintbrush paint below the horizon line with the 'soil paint' and allow to dry.

5 For the roots, cut a number of different lengths of string, of varying thickness if possible. Water down a little PVA glue in a bowl and, one at a time, dip each piece of string into the glue. Form the roots by laying each piece of string on the soil from the horizon line downwards.

6 For the plant stems cut strips of green paper or use art straws and apply above the horizon line. For the leaves cut several simple leaf shapes from the fabric or paper and glue on, allowing them to curl up for a three-dimensional effect.

Dreamtime Art

Dreamtime Art is the name of art produced by or in the style of the indigenous Australians. Traditionally they served as maps of the surrounding landscape which contained important information about sources of water, animals, and where communities lived. Use the internet to research these symbols and use them to create your own map showing the journey from home to school in earthy colours.

Resources

- Clay
- Sand
- PVA glue
- Poster paints in earthy colours – burnt umber, sienna, orange, ochre, white and black
- Cardboard panels from a cardboard box – approximately A3 size
- Pencils (or fingertips)

Approach

1 Put a handful of clay, a tablespoon or two of sand, and a dollop of PVA glue onto the board and mix the three materials together, smearing the mixture across the board.

2 Squeeze on the chosen paint colour and lightly smear it across the textured mixture on the board. Put it aside to dry.

3 Add all of the colours to the pallet, and dip the flat end of a pencil or fingertips into the paint and make dot formations on the dry board.

4 A small circle signifies the home, a large circle signifies the school, and curved lines joining the two show the child's journey to school. Children can refer to the Dreamtime symbols available to elaborate their work, and should be encouraged to talk through their work.

A Show of Shadows

These shadow puppet theatres give children the opportunity to create new stories or adventure worlds of their own and retell favourite traditional fairy tales. Using light and colour in a container, these theatre boxes have a magical quality that work even better in the dark, and demonstrate how silhouettes are formed.

Resources

- Black acrylic paint
- Black sugar paper
- Cutting knife (teacher use only)
- Tissue paper in a variety of colours
- A laminator and A4 laminating pouches
- Plastic straws or sticks
- Small battery-operated light
- Sticky tape
- Glue sticks
- Shoebox with lid

Approach

1 Paint the outside and lid of the box with black paint, and the inside white and leave it to dry.

2 Using a cutting knife (teacher to do this) cut a large rectangular panel from the box lid, so that it resembles a TV screen.

3 Tear or cut pieces of tissue paper and carefully glue them together in a patchwork-style to the same size as the box lid.

4 Once dry, laminate the delicate tissue paper panel. This will strengthen, flatten and intensify the colours of the panel.

5 Turn over the box lid and trim the laminated panel to size and tape it into position.

6 Draw two simple pictures on black sugar paper. This will be the theatre set. For example, trees for an outdoor setting, a castle for a traditional fairy tale, or seaweed for an underwater scene. Cut out the pictures and stick them on the inside of the colour panel.

7 Draw the main character (the moving component) on black sugar paper or card and attach a straw with sticky tape.

8 Cut a slit the length of the box across the top with the cutting knife (teacher to do this).

9 Turn on the light and place it inside the box, put the lid back on the box and stand upright on a flat surface. Put the moving component through the slot and begin the puppet show!

The Power of the Sun

The main resource for this project is strong sunlight! A nice sunny day is perfect for getting some quick and dramatic results on paper. The focus is on the Sun's ability to bleach and fade colour from artificially coloured materials such as paper and some plastics. There is also an important link to the properties of sunlight and its ability to provide light, heat and power.

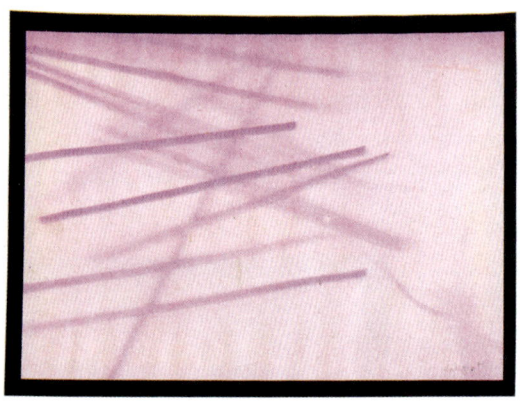
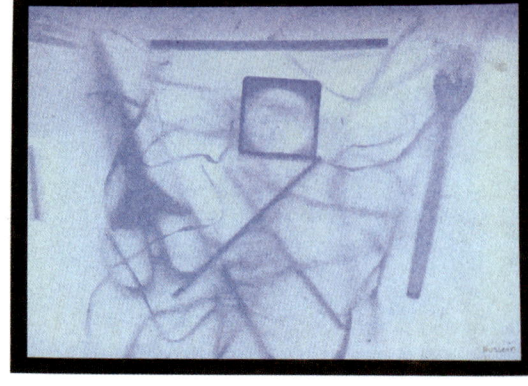

Resources

- A4 sugar paper in strong colours
- A4 white paper
- A collection of objects, for example, string, scissors, raffia, lolly sticks, art straws, pebbles
- Masking tape

Approach

Picture with stencils

1 Cut shapes from the piece of white paper and place over the piece of coloured sugar paper. Secure the two pieces together with masking tape.
2 Place the picture in strong sunlight either on the ground if practical, or taped to a window that catches the Sun.

3 Check at hourly intervals if the paper is outside, and at daily intervals if pinned to the window, lifting a corner to see how much the paper has faded.

Picture with objects

1 Place the paper outside at the middle of the day when the Sun is at its strongest. Or place the paper on a windowsill that gets lots of sunlight. Arrange a selection of objects on top of the paper and return in an hour or two to check results.

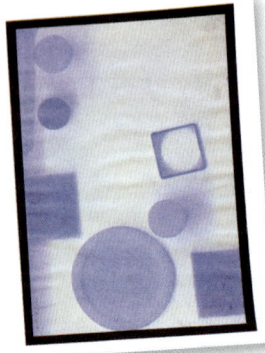

Sunrise, Sunset

These small-scale delicate works look at the natural phenomenon of the Sun rising in the East and setting in the West. The Victorian artist JMW Turner (1775–1851) and the impressionist Claude Monet (1840–1926) painted some beautiful sunrises and sunsets on the River Thames in London and towards Kent. They looked in detail at the many colours that appear in the sky at these two particular times of day. Though based on the Thames, any city or townscape can be superimposed on the watercolour background and adapted to the place where you live.

Resources

- A4 cartridge paper
- Watercolours
- Pink, yellow and orange paper (fluorescent if possible)
- Glue sticks
- Black paper

Approach

1 Using a large paintbrush spread water across the paper and then lightly brush sunset colours of pink, yellow and red onto the damp surface, taking time to let the colours naturally spread across the paper.

2 Once dry, use the pink, yellow and orange papers to create a hazy, low-lying Sun. Place the Sun to the right of the picture to show the sunrise, and to the left of the picture to show the sunset.

3 Cut out buildings using the black paper and stick them on the horizon line.

The Four Seasons

Using identical images makes this project a great way to explore different materials in order to alter the same image in different and radical ways. Including an image of a person posing as if in the rain or Sun brings an added element of fun to each collage in the series. To create the image, take pictures of the local environment or of the school grounds.

Resources

- A4 sugar paper in strong colours
- A collection of objects, for example, string, scissors, raffia, lolly sticks, art straws, pebbles
- Masking tape
- Glue sticks
- A4 white paper

Approach

1 **Spring:** Using the pastels draw a Sun in the sky of the first photocopy.
2 Turn the picture over and draw a series of diagonal lines approximately 2cm apart on the back of the picture, from one corner to the other and in the same direction.
3 Cut along the lines and lay the strips on the black paper to recreate the image. Separate the strips, different widths apart, to create a fractured image. Once arranged, stick the strips to the paper with glue.
4 **Summer:** Look at images by the artist van Gogh (1853–1890). Using oil pastels with short quick marks, build up layers of colour to create a sunny scene.
5 **Autumn:** Fold the tissue paper three or four times and cut out leaf shapes. Fold along the centre and draw veins on one side.
6 Glue the creased leaves onto the image to give a textured effect. Build up to create a flurry of leaves.
7 **Winter:** Tear small amounts of cotton wool and twist between the fingers. Apply glue to all the surfaces that snow would naturally fall on and place the cotton wool. Take care not to touch the glue with your fingers, as this can make it difficult to handle the cotton wool.

Volcanoes

Volcanoes are always popular with children because of their exciting explosions and dramatic bright orange lava. This project aims to recreate some of the action of an exploding volcano with a handmade and hand-held device. The landscape surrounding your volcano can be adapted to show different environments, such as fertile farmland or recreating the city of Pompeii after Mount Vesuvius covered the entire city in ash and pumice in the year 79AD. A study of Pompeii would form an excellent basis for this project.

Resources

- Flour
- Cardboard
- String (optional)
- Stapler (teacher use only)
- Acrylic or poster paint – dark green, light green, grey, fluorescent orange
- Polystyrene chunks or sponge pieces
- Art straws – red, orange and yellow
- Cutting knife (teacher use only)

Approach

1 Draw a small circle at the centre of the paper plate, cut a line from the edge of the plate to the centre and then cut out the small circle at the centre.
2 Overlap the two open sides into a cone shape, and staple into position (teacher to do this).
3 To create the rocky volcano, break the polystyrene or sponge into rough chunks. Using generous amounts of PVA glue, stick the pieces onto the cone leaving a small space between each one. Allow them to dry.
4 Cut a piece of card to a preferred shape and size, and using the cutting knife (teacher to do this) cut a hole which is smaller than the cone. Apply a generous amount of PVA glue to the bottom rim of the cone and place over the hole on the larger piece of cardboard and allow it to dry. You can place a book on top of the cone to help join the cone and board completely whilst drying.

5 Once dry, paint the board with both shades of the green paint to create a landscape, and paint the volcano grey and allow it to dry.
6 To make the lava flow thick and slow down the sides of the volcano, mix a small amount of the fluorescent paint with the PVA glue and a little flour into a paste. Pour slowly around the rim of the volcano and allow it to trickle down.
7 For spraying lava bunch together several art straws and tape them together at the bottom.
8 To secure the two pieces together make a hole at the edge of the board and feed the string through the hole and secure with a knot. Tie the other end of the string around the bottom of the straws and push the straws into the volcano. Poke the straws through the top of the crater to demonstrate the explosion!

A Time for Reflection

Create a mini gallery and optical space using light, colour and reflection. Through the use of collage, drawing and shape-cutting you can create a variety of interesting surfaces that are reflected in new and interesting ways in the mirrors. For inspiration look at the artwork of M.C. Escher (1898–1972) whose work is about illusion, and the contemporary artist Anish Kapoor (1954–) who often uses mirrors and distortion in his work.

Resources

- Glue sticks
- Oil pastels
- Sticky tape
- Scissors
- Coloured paper
- Plastic concave/convex mirrors 10cm x 10cm, available from good school suppliers (two per person)
- White flexible card 10cm x 10cm (three per person)

Approach

1 Start by looking at the effects of concave and convex mirrors and see how they distort the images they reflect.
2 On the first of the three pieces of card, cut small shapes from the edge of the square and in the centre (if possible), taking care not to cut away an entire edge.
3 On the second piece of card glue a variety of coloured paper pieces in small squares, strips or in shapes of choice.
4 The third square will be the base. Use oil pastels on it to make a colourful abstract surface.
5 Using sticky tape join the first two squares together with all the colour appearing on the inside. They can be taped together whilst flat on a table, leave a millimetre gap for folding. Then add the third square as the base.
6 Lastly tape the two plastic mirrors together to make a corner. Note: the most visually effective combination is to position the concave mirror opposite the white square with cut shapes, and the convex mirror opposite the collaged square. The oil pastel square remains the base square.

Reflective Colours

This is a simple collage project that uses highly contrasting colours to illustrate absorbent and reflective surfaces. Colours that reflect light include white, light colours containing white (such as pale blue) and fluorescent colours. Colours that absorb light are very dark, black being the most absorbent of all. As a source for imagery, discuss light sources that occur in the dark, for example, cat's-eyes on a motorway, runway lights and street lights. These ideas can form the basis for a simplified pattern for your picture. Look at artwork by Bridget Riley (1931–) for optical effects.

Resources

- A3 black sugar paper
- Variety of fluorescent colours such as orange, pink, green and yellow
- Glue sticks

Approach

1 To work out a pattern or image, draw several rectangles on a piece of paper. Within each rectangle sketch an idea, taking inspiration from an image, such as a street lit up at night.

2 Cut strips and shapes from the fluorescent paper and arrange them on the black paper. Symmetry and pattern designs work particularly well with these strong colours.

3 Once you have your arrangement stick the coloured pieces down with glue.

Growing Sunflowers

This project combines two areas; it is inspired by van Gogh's (1853–1890) *Sunflowers* series (1888–1889) and explores plant structures. This simple model shows the main components of a growing plant – the stem, the roots, the leaves and flower – using different materials to create them.

Resources

- Cardboard
- Green fabric, paper or card
- Green art straws
- Yellow and gold paper
- Sunflower seeds
- PVA glue
- White string or shoelaces

Approach

1 Cut several strips from the cardboard, approximately 45cm x 2cm. Cut a small square approximately 6cm x 6cm and attach one long strip of the cardboard to the square using the PVA glue, lying flat to dry completely.

2 Glue three green art straws or green paper strips onto the long strip of card, starting just below the square

panel. Leave a small space at the bottom – this will become the stem.

3 Glue four leaves to either side of the stem, they will hang off the sides like a real plant. If they are fabric you may want to cut them out in advance for younger children.

4 Cut out petals from the yellow and gold paper. You can use crinkle-effect scissors but plain scissors are fine and the petals can be as large as is manageable! Stick one end of each petal to the centre of the square at the top of the stem.

5 Spread a generous amount of PVA glue onto the centre of the square and sprinkle a handful of sunflower seeds over it, pressing the seeds down so they are secure.

6 For the roots, tie two shoelaces together or knot together some string lengths. Glue in place at the small space on the card just below the green art straws or paper.

7 Leave flat to dry completely and then display together as collection of growing sunflowers.

Growing Pictures

Create your own shapes using cress and watch your picture grow in a matter of days. This project is based on making shapes in paper, a bit like large cookie cutters. Cress seeds will germinate and grow in whatever shape you choose. The project takes inspiration from artists who work with nature, such as British artists Andy Goldsworthy (1956–) and Richard Long (1945–), and American artist Jeff Koons' (1955–) sculpture, *Puppy* (1992) outside The Guggenheim Museum in Spain (pictured below). This indoor plant project can be done at any time of the year and is not dependent on strong sunlight.

Resources

- Plastic trays
- Cotton wool pleat
- Cress seeds
- Sticky tape
- Paper or flexible card
- Water
- Scissors

Approach

1 Cut strips of paper or card approximately 25mm wide. Bend into your chosen shape and tape into position. For example, a circle, triangle, square or diamond.
2 Pour a little water into the tray. Place the cotton wool in the tray allowing it to absorb the water. Place the paper shape on the cotton wool surface.
3 Scatter the seeds on the moist cotton wool within the shape, remove the paper silhouette and cover with a sheet of paper until the shoots reach 25mm. For a sharper outline of the shape, leave the paper perimeter in place whilst the cress shoots grow.

4 Keep the cotton wool moist as necessary. Once the shoots have grown to 25mm remove the sheet of paper and allow them to flourish in the daylight.
- Option: photograph the cress when at its peak as a memory of the work.

Minibeasts

This project uses a plastic magnifying sheet (inexpensive and available to purchase online) to create a scientific viewfinder effect. Look at the habitats of minibeasts and their ability to camouflage themselves, or to be noticed by their very bright colours (ladybirds, for example). Start by drawing the insect to focus attention on the detail of the minibeast. Look at how many legs it has, its colouring and shape. Describe its habitat, consider what kind of environment it likes and how it moves around.

Resources

- Green, brown or yellow art straws
- Red art straws or paper
- Green fabric or paper
- Buttons
- Raffia, string or wool
- Sticky tape
- Standard plastic magnifying sheet 27cm x 19.5cm
- Stiff cardboard panel approximately 18cm x 19cm
- PVA glue

Approach

1 Cut the art straws in half, and using any colour of choice stick the straws vertically across the width of the cardboard.
2 Once dry, trim the straws so they are the same width as the cardboard panel.
3 Cut two leaf shapes from the green fabric or paper.

Apply a blob of glue to the panel and lay each leaf down, taking care not to press it flat so that it has a slight curl to it.

4 To make the minibeast use materials appropriate to the insects you are studying. Minibeasts that have hard bodies, like an ant or a beetle, can be made with large and small buttons. The art straws can be cut into short lengths and assembled one at a time with PVA glue to make minibeasts.

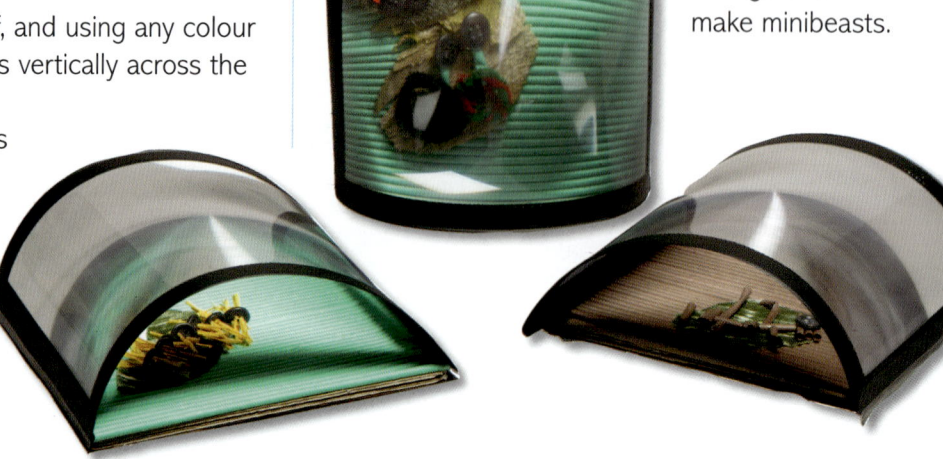

The Butterfly Life Cycle

This is a three-dimensional diagram of the life cycle of a butterfly, from caterpillar to chrysalis to butterfly and back to the egg. The models shown here have been largely made with paper.

Resources

- Paper plate
- Diluted brown paint
- Nail brush or stiff brush
- Tissue paper – green, brown, black, white, yellow
- PVA glue
- Coloured paper
- Buttons
- Raffia or string
- A selection of materials – pipe-cleaners, coloured sticks as available
- Stapler (teacher use only)

Approach

Wood base

To create a wood-effect on the paper plate dip a nail brush into the diluted brown paint and drag it across the back of the plate. After drying cut a line from the edge to the centre of the plate. Overlap the two edges to form a dome and staple into position (teacher to do this).

Caterpillar

Cut a straight or slightly curved piece of coloured card. Take several buttons and poke short pieces of raffia or string through each hole. Put a generous amount of PVA glue onto the paper strip and place each button on top. Once dry stick onto the plate.

Chrysalis

Gently roll a piece of black or brown tissue paper into a small sausage shape, then wrap a single layer of green tissue paper around it and glue into place. Bend to curve it a little and stick it onto the plate.

Butterfly

Using the coloured card, fold a piece in half and cut out a wing shape. Open the card to reveal a symmetrical butterfly. Add detail using drawing materials, stickers or with any material of choice. A pipe-cleaner or stick can be used for the body and antennae. Once dry, apply glue to the butterfly and add to the plate.

Egg

Roll a little white tissue paper into an egg shape and stick it onto the plate.

Roots and Shoots

This project is a 3D model of a growing plant showing the roots, stem and leaves. The box represents the soil and can be embellished with minibeasts as required.

Resources

- Fine modelling wire
- Green paper
- White acrylic paint
- Glue sticks
- Shoeboxes
- Brown poster paint
- Blu-Tack®
- Coloured pencils or paper

Approach

1 Twist several strands of modelling wire together in the centre. Fan out the top of the strands for the plant stems and fan out the lower strands for the roots. Aim to make the model self-supporting by standing the roots on the table.

2 Fold a small piece of green paper and cut out a double-sided leaf. Apply glue across the bottom of the leaf and fold it around the wire stem. Build up the stems with lots of leaves.

3 Paint the lower half of the wires (the roots) with white acrylic paint and allow them to dry.

4 Meanwhile, paint the inside of the box with brown paint, varying the shades adding a little yellow or a lighter shade of brown.

5 Once dry cut out small worms, ants and snails using coloured paper or hand-drawn underground insects, and stick them inside the box.

6 To insert the plant into the box, cut a narrow slit at the top of the box and slide the plant in and secure the underside with Blu-Tack®.

Leaf Veins and Patterns

These bold pictures look at the veins and patterns on different leaf types. Start by looking at varieties of tree and the leaves they produce. If this project is being done in autumn gather a selection of leaves and begin the project with some observational drawing. Otherwise use pictures as your resource. Draught-excluding tape is used to recreate the leaves on a larger scale, it is cheap and available from most DIY stores.

Resources

- A1 shiny white paper or cartridge paper
- A1 black sugar paper
- Draught-excluding tape
- Acrylic or poster paint in different shades of green
- Glue sticks

Approach

1 Cut a piece of draught-excluding tape approximately 10cm long and gently place it on the paper, taking care not to press it down. It will be removed later, so it shouldn't be stuck down yet.
2 Continue to cut lengths between 10 and 15cm to form the veins and a longer piece for the outside edge of each leaf. Make two to three leaves, placing them centrally on the paper.
3 Paint the entire sheet of paper, and around the tape with different shades of green.
4 When the sheet is dry, very carefully remove the tape and discard the pieces.
5 Draw a large tree on the black paper and cut it out. Discard the cut-out shape and rub glue around the edge of the tree shape and at the edges of the black paper. Carefully lay the black paper over the green picture and press it flat. Trim it with a paper cutter as necessary.

Rainforest Habitat

Research tropical rainforests and explore the varieties of plant and animal life that you would find there. This project is adaptable to many different habitat settings.

Resources

- Pictures of tropical animals
- Oil pastels
- Different green printing inks
- Trays
- Pencils
- A2 white flexible card or paper
- A4 or smaller poly board
- Rollers
- Glue sticks

Approach

1 Colour the image of the tropical animal using oil pastels then cut it out. Put it to one side.
2 On a piece of poly board, draw the detailed tropical leaf as large as possible. Cut this out to make a printing block.

3 Set up trays of printing inks, each containing a different shade of green. For a dark green, mix green with a little dark blue, for a lime green mix green with yellow.
4 Print four to six leaves in different shades and allow them to dry. At some point the colours will mix together whilst inking up, this will increase the richness of the colours as they combine and there is no need to wash the block whilst working.
5 Once dry, cut out the printed leaf and fold in half along the centre, then lay it flat.
6 Glue the animal pictures in the centre of the card or paper and attach the leaves by gluing along the fold so that they are not completely attached. Use all the leaf prints to make a dense rainforest environment.

Seaside Habitat

As a starting point look at the variety of species of fish and marine life that exist off the shores of your country of choice. The pictures here include British marine life such as shore crabs, hermit crabs, cod and John Dory. These creatures are very good at camouflaging themselves and the colour scheme used here reflects this. The project is adaptable to many different habitat settings. Use the various features of the habitats to make printing blocks.

Resources

- Pictures of marine life
- Real shells (or pictures of shells)
- Oil pastels
- Orange, yellow, ochre and brown printing inks
- A4 or smaller poly board
- Pencils
- A2 white flexible card or paper
- Trays
- Rollers
- Glue sticks

Approach

1 Colour the image of marine life using oil pastels and cut it out. Put it to one side.
2 If possible, spend some time doing some observational drawings of real shells, looking closely at the lines and markings that make them distinct. Then, on a piece of poly board, draw the shells in detail as large as possible, creating a rock pool of shells and pebbles.

3 Set up trays containing the various printing inks.
4 Print four shells in the different colours and allow them to dry. The colours will mix together whilst inking up. This will increase the richness of the colours as they combine and there is no need to wash the block whilst working.
5 Once dry, glue each of the four prints horizontally onto the card or paper. For a three-dimensional effect put a line of glue at the top and bottom of each print and allow it to ripple before securing in place.
6 Glue the chosen marine life image in the centre of four prints, and ripple before gluing if preferred.

Busy Bees

This project demonstrates insect pollination by recreating a garden heavy with pollen and nectar in the summertime, and a bee visiting two flowers. The flower heads include a stigma and stamen. They are a little fiddly to put together; working in pairs may help to overcome this.

Resources

- Plasticine®
- Cotton wool
- Orange or yellow powder paint
- Crêpe paper in various colours
- Fine modelling wire
- Yellow and green art straws
- Glue sticks
- Raffia, string or elastic bands
- Yellow and white paper
- Black pipe-cleaners
- Small buttons
- A small cardboard panel
- Green tissue paper or paint

Approach

Flowers

1 Take a small piece of cotton wool – this is the flower head – and poke a short length of a green art straw into the centre; this is the stigma. Glue short lengths of yellow art straws onto the outside edge of the cotton wool, these will form the stamens.

2 Fold a piece of crêpe paper several times and cut out a petal shape; this will produce several petals. Glue the petals together at the base, and glue them around the edge of the cotton-wool flower head. To make them secure tie some string, raffia or an elastic band around the bottom.

3 Take a length of modelling wire and thread through the flower head knot, and pull through to make each length even. Twist the wire together to secure.

Bee

1 Roll a piece of yellow paper into a tube-shape and wrap the black pipe-cleaner around it, securing in a twist at one end. Thread a button onto each piece of pipe-cleaner that is sticking out; these are the bee's eyes.

2 Cut another pipe-cleaner to make the hairy legs.

3 Add two paper wings, secure by tucking them under the pipe-cleaner.

4 Wrap a piece of modelling wire around the bee's body, twisting it together underneath the bee to secure it.

Assembly

1 Put three small lumps of Plasticine® onto the board and cover using PVA glue and then green tissue paper.

2 Poke the modelling wire from the bee into the central lump of Plasticine®. Stick the flowers in the two remaining lumps of Plasticine®.

3 For the pollen, sprinkle half a teaspoon of powder paint onto one of the flowers, and gently bounce the bee onto the flower. His hairy legs should pick up the pollen and when you gently move the bee to the next flower, some of the pollen will be deposited, as in nature.

Seeds Airborne

These floaty, wispy paintings capture the journey of airborne seeds being dispersed as they are carried on the wind. These pictures are based on dandelion and sycamore seeds, and use paint in a textured hand-applied technique.

Resources

- A3 white cartridge paper
- Poster paints
- Sponges, tissue or newsprint paper
- Flexible card
- Masking tape

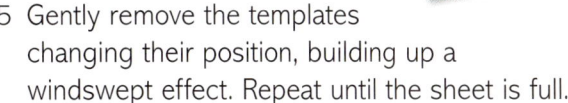

Approach

1 Make several templates of the chosen seed shape and place them on the paper at random intervals. They can be dropped onto the paper, mimicking the windswept journey of seeds.

2 Fold a small piece of masking tape into a loop and use it to stick each template to the paper.

3 Choose two similar colours (for example, pink and purple or yellow and green) and white for your pallet.

4 Dip the sponge or tissue lightly into the paint, taking care not to make it too wet. When each template is in position, lightly dapple over and around the template creating a wispy effect keeping the tissue quite dry. Work in a small amount of the other colour. Add a little white to the colour work to lighten the shades.

5 Gently remove the templates changing their position, building up a windswept effect. Repeat until the sheet is full.

6 The templates can be glued to the paper for a positive/negative effect.

61

Jurassic Jewellery

This costume project allows the children to dress up like dinosaurs and invent their own colour scheme and scale shapes. It can also be adapted and is great for assemblies or carnival time. Start by looking at the scaly skins of reptiles, such as snakes and crocodiles, which come in many pattern formations.

Resources

- Cardboard
- Acrylic paints
- Flexible card
- PVA glue
- Elastic or string
- Cutting knife (teacher use only)

Approach

T-rex arms

1 Cut two rectangular pieces of flexible card approximately A4 size and paint with any colours the children choose, blending and stippling to give texture. Put aside to dry.

2 Measure the panel length against the child's arm and if it is slightly too long make a fold at the wrist and cut vertical lines up to the fold. This will fan out and cover the hand.

3 For the scales, take a small panel of the thick cardboard and paint with any colours the children choose. Once dry, cut it into small shapes. Off-cuts are useful for further interesting shapes. When the arm panel and the shapes are dry, glue them on using PVA glue, and allow them to dry.

4 Once dry, carefully bend the arm panel into a tubular shape, position it just under the child's elbow, allowing the arm to bend. At the back make two holes at the top and bottom of the piece where they join and feed the elastic or string through those holes, tying them in a knot, so that the children can put the arm on and take it off easily.

Lizardy legs

1 Cut two rectangular pieces of flexible card. Paint with any colours the children choose, blending and stippling to give it texture. Put aside to dry.

2 Measure the panel length against the child's lower leg. Make a fold at the ankle and cut vertical lines up to the fold. This will fan out and cover the front of the foot. Trim these strips into claws.

3 For the scales, follow the instructions as given in the 'T-rex arms' details.

4 Once dry, carefully bend into a tubular shape, position just under the knee allowing the leg to bend. Secure with elastic or string as explained in the 'T-rex arms' details.

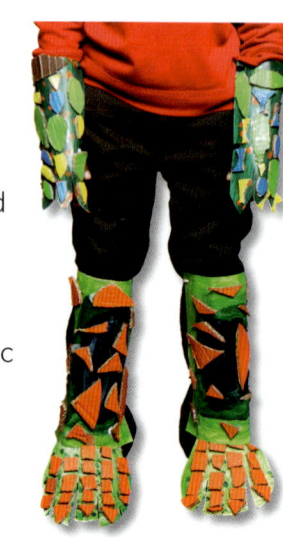

Mega-Stegosaurus and Dimetrodon

These giant-scale dinosaur costumes are made using simple materials. The children can create their own scaly designs and colour schemes and assemble the costumes to wear; perfect for assemblies or carnival time.

Resources

- Cardboard
- Acrylic paint
- Flexible card
- PVA glue
- String
- Cutting knife (teacher use only)
- Art straws

Approach

Mega-Stegosaurus

1 Draw large plates on cardboard. Cut them out with the cutting knife (teacher to do this). Keep the excess scraps as they will be useful for extra patterning.
2 Paint the pieces in any chosen colours on both sides and allow to dry.
3 To make the costume structure, cut two strips of flexible card approximately 80cm in length, but make it smaller or larger according to the length of the wearer's back. Fold each strip in half lengthways and glue the plates to the inside with the PVA glue.
4 Once dry, paint any unpainted areas of the flexible parts that are showing, blending the colour with that of the existing plates. The spaces in between each plate can be cut to shape to enhance the piece.
5 Attach string loops to the top and put arms through these loops to wear the piece.

Dimetrodon

1 Draw the lizard-like fin on cardboard. Cut it out with the cutting knife (teacher to do this).
2 Paint the piece in any chosen colours on both sides and allow it to dry.
3 To make the costume structure, cut two strips of flexible card approximately 70cm in length, but adjust as appropriate. Fold each strip in half lengthways and glue the fin to the inside with PVA and, using more PVA glue, press it shut to stick together and dry.
4 Once dry, paint any unpainted areas of the flexible parts that are showing, blending the colour with that of the existing fin.
5 Attach string loops to the top of the piece to use as arms holes.

Deep-Sea World

Tropical fish and colourful coral are the focus for this project that examines the creatures of the oceans. There are some fantastic images of tropical fish and endangered corals on the internet that provide a good stimulus for children to create their own sea world. This project uses plastic magnifying sheets, which are inexpensive and available to purchase online.

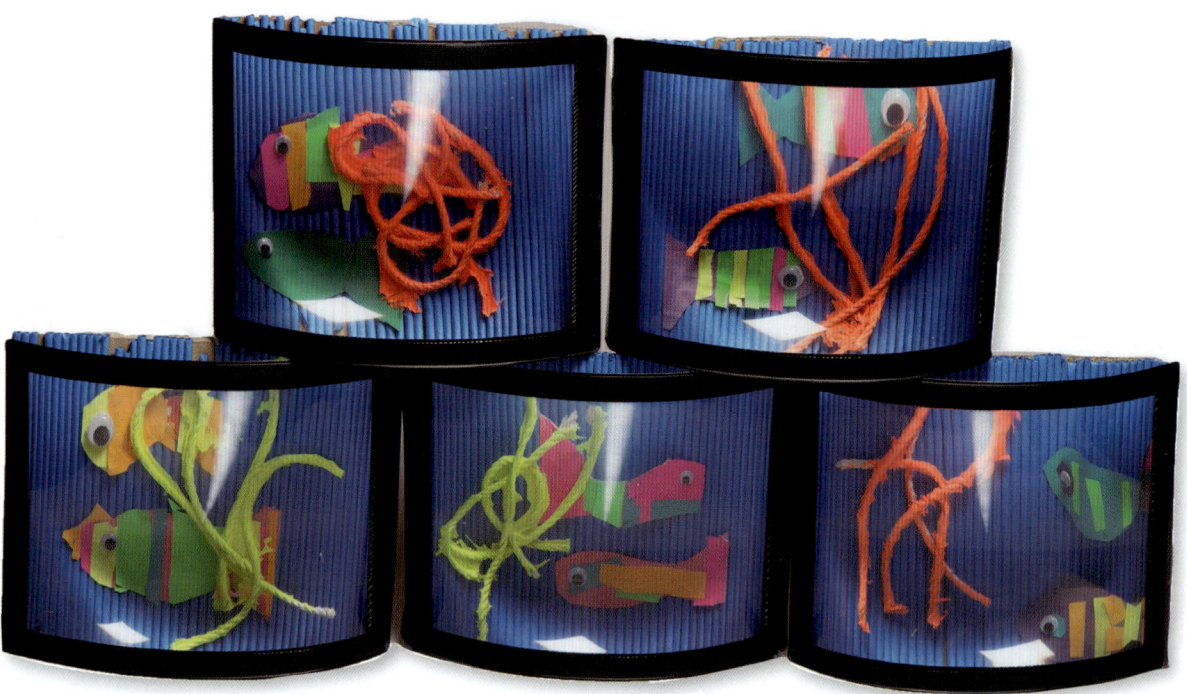

Resources

- Blue art straws
- Cardboard 19cm x 25cm
- Standard plastic magnifying sheet 27cm x 19.5cm
- Brightly coloured paper
- PVA glue
- Flour
- Fluorescent paint
- String
- Cling wrap or a plastic sheet
- Tray or board
- Googly eyes or buttons
- Sticky tape

Approach

1 Cut the blue art straws in half, and stick the straws vertically on the cardboard panel. Once dry, trim any overhanging pieces of straw.
2 Use the colourful paper to create two colourful fish and glue to the blue straws. Gently bend the fish to give them a 3D look.
3 To make the coral, mix the fluorescent paint with a little PVA glue and a little flour, cut five or six pieces of string of varying length, and dip each one in the paint mixture and lay onto the plastic sheet (a cut-up plastic bag will do). Continue to create a lattice effect and allow to dry on the tray.
4 When the coral is dry peel back from the cling wrap and glue onto the blue straw panel.
5 Attach the magnifying sheet with sticky tape to both sides of the panel.

Creatures from the Deep

New species of marine life are being discovered all the time, and this project looks at the fascinating squid and invertebrates that inhabit the deepest part of the ocean. As the light does not reach the furthest depths, many creatures make their own light by glowing in the dark and having fluorescent colours. Their strange shapes, big eyes, tentacles and transparent membranes are a good starting point for the children to create their own creature of the deep. As many have dappled colouring, look at the work of Georges-Pierre Seurat (1859–1891) who used a technique called pointillism whilst painting.

Resources

- White flexible card or paper
- Acrylic or poster paint
- Metallic paint
- PVA glue
- Cardboard panel
- Coloured acetate (optional)
- Sticky tape

Approach

1 Draw a large squid on the flexible card and apply a variety of colours in dots, using a small paintbrush.

2 Once the paint is dry, cut the squid out and glue it onto the metallic paper. Cut it out again, but leave an edge of 1cm all the way round. Add details using metallic paper, such as shiny eyes and different coloured tentacles.

3 For the background, paint a panorama of the deep ocean using different shades of blue. Put it aside to dry.

4 Once the background is dry, gently bend the squid to create a 3D effect and then attach it to the background using generous amounts of PVA glue.

Giant Mammals – Footprints

Begin by looking for images of large mammal footprints on the internet. Because they live in diverse habitats there are some interesting surface textures to be found including snow, dried and cracked river beds, muddy soil, etc. The aim of the activity is to recreate these surfaces using colour and textured materials.

Resources

- A piece of canvas or J-cloth™
- Poly block or cardboard
- Sponges or large paint brushes
- Air-dry clay
- Modelling tools
- Cardboard 30cm x 45cm
- Scissors
- PVA glue
- Poster paints

Approach

1 To make the footprint, begin by practising drawing the shape on paper. When ready, draw a left and a right footprint approximately 4cm x 5cm on the poly block or cardboard and cut them out. Put to one side.

2 Place a J-cloth™ or piece of canvas on top of the cardboard piece. Place the clay on top of the material and roll it out to a depth no thinner than 2cm.

3 Mark the clay using the modelling tools to recreate the texture of the ground.

4 Take the footprints 'for a walk' across the clay surface mimicking the appropriate animal. Press down sufficiently to leave a footprint in the clay. Remove the cardboard footprints carefully after each step.

5 Once dry, apply the paint with a stippling action using either a sponge or the tip of a thick brush. Apply the colour patchily at first and build up to resemble the soil surface.

6 Paint the footprints with a smaller brush to create a smooth surface in black or an appropriate colour.

7 To add shine to the footprints, add a little water to the PVA glue and brush on two coats.

Giant Mammals - Sea Mammals

The inspiration for this project is the later work of Pablo Picasso (1881–1973). Strong colours and structured lines have transformed these sea mammals into striking wearable artworks. Look at the paintings Picasso is most famous for, such as *Guernica* (1937), *Marie-Thérèse Walter* (1937) and *Three Musicians* (1921). Sometimes the faces that are in profile have two eyes on the same side and most have very bright colours used in sections. These animal hats are great for assemblies, a summer carnival or school plays, and can be adapted to any theme.

Resources

- Four flexible card strips 2–3cm in width, approximately 70cm in length
- Flexible card or heavy paper in large sheets
- Acrylic paint in a variety of colours
- Drawing pencils
- Scissors
- Stapler (teacher use only)

Approach

1 Using one of the strips; cut a headband to size and staple securely (teacher to do this). Make the internal structure by crossing a further two strips over the head from ear to ear, and front to back. Tape in position and staple to secure (teacher to do this). Put to one side.

2 Draw the animal in profile on the large sheet of flexible card or paper and cut it out. Place on top of another piece of paper and draw around it and cut it out. There will now be two identical images that make both sides of the sea mammal.

3 Draw the detail of the sea mammal on both of the cut-out shapes in the style of Picasso. Both sides should be a little different from one another, making it interesting to look at as the wearer turns around.

4 Paint each section in a different bright colour.

5 Once dry, put the two halves of the hat together and staple at the top edge approximately 8cm apart.

6 The teacher will need to do this next step. Put the helmet/ internal structure on the artist's head and position the two stapled sides of the 'hat' over the helmet.

7 Adjust the hat until the headband cannot be seen, but the wearer still has good visibility. Using a finger and thumb hold the hat to the helmet and remove carefully. Staple the two parts (teacher to do this) into position (once off the wearer's head). Reposition and repeat as necessary until the hat is secure and wearable.

8 Consider adapting for other sea creatures such as sharks!

Food Chain Model

This is an interactive and illustrative model showing the food chain in four different habitats. A 2D option is also possible. Begin by looking at food chains in four different environments (woodland, the Arctic, African plains and the ocean). A good starting point is to look at images of each life form in the food chain for the children to work from.

Resources

- Two poly block squares approximately 15cm x 15cm
- Printing inks
- Four empty boxes in graduating sizes
- Coloured paper
- Glue sticks
- Large sheets of cartridge paper

Approach

1 Begin with the print-making activity. Draw the chosen image on the poly block tile and carefully cut it out, gluing it onto the second tile using PVA glue. Use plenty of PVA glue as the tile will make many prints so it must be robust. Any additional details can be added with extra cut pieces of poly block and glued on. Finish using a sharp pencil for line work.

2 Once dry, select appropriate colours for the life form (for example, shades of green for a frog, shades of blue for Arctic habitats). Starting in one corner work methodically across the sheet repeating the image to fill the sheet. If pieces become detached whilst working, staple them back on (teacher to do this).

3 Cut the cardboard boxes down to the same height and where necessary secure the structure using sticky tape on the flaps and openings.

4 Once the printed sheets are dry, select the four images for the bottom of the food chain and cover each side of the box with the chosen sheet, like wrapping a gift. The top part of the box can be covered with a triangular flap, with all four points meeting in the centre.

5 Continue working upwards through each tier, making sure that the sides correspond and are in the correct order. Trim off any excess paper where necessary.

6 To neaten the corners add a coloured strip to the edges of the box and strips across the top of each box to separate the pictures. Place coloured strips in an 'x' across the top of each box.

7 Lastly, label each life form.

- Option: create a 2D pyramid for a wall display by decorating and labelling each section in the same way as the 3D model.

Camouflage

These pictures begin as abstract pieces looking at well-camouflaged animals in their natural habitat. They provide the opportunity to look at the patterns found in places such as an African plain, a leafy forest or an underwater scene. The animal silhouettes add a presence that stands out and blends in at the same time. Start with researching those animals large and small, on land and in water, who are masters of disguise!

Resources

- A2 cartridge paper
- Acrylic or poster paints
- Brushes in a variety of sizes
- Animal pictures
- Black, white or coloured paper
- PVA glue

2 Using a large brush for bigger areas, and smaller brushes for detailed areas, begin to layer on, smudge and apply paint in a textured way, building up the pattern of grass, mud swamps or animal patterning as chosen.

3 Once dry, draw a silhouette of the animal on the black paper and cut it out.

4 Cut out the middle of the animal silhouette leaving a 5mm–1cm edge all the way around.

5 Position the animal silhouette onto the painting. Use a thin layer of PVA glue around the edge of the silhouette and press it flat.

Approach

1 Look closely at the colour and texture of a camouflaged animal in its natural habitat. Select colours from the pallet that closely match those in the image. It is also a good opportunity for the children to try and mix different colours and shades.

Cells in Action

This project demonstrates cell division and is a good illustration of how bacteria can spread rapidly, through the simple model of a flip book. The front and back cover designs can be an opportunity to use drawing and design programmes on the computer.

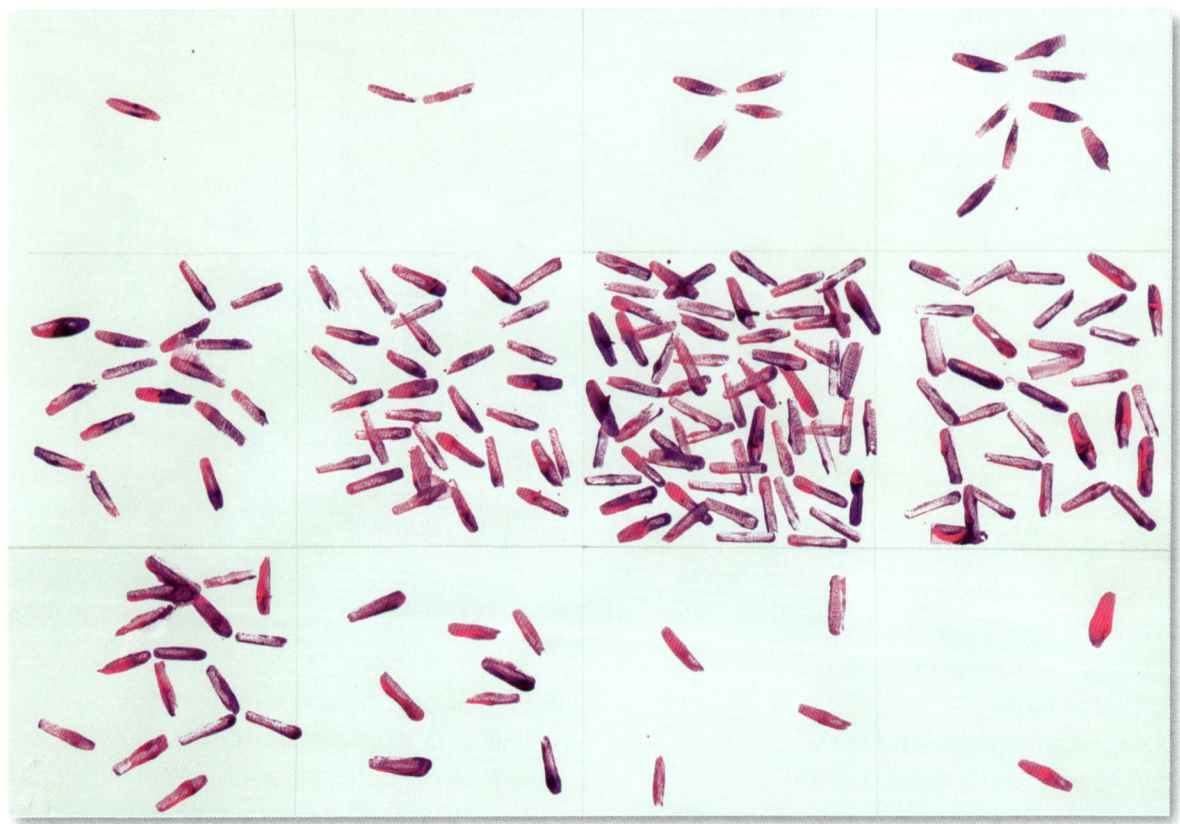

Resources

- A3 white paper
- Drawing pencil
- Coloured pencils or felt pens
- A ruler
- Poly board
- Poster paint
- Stapler
- Images of bacteria reproducing
- Coloured paper

Approach

1 Within a 10cm x 8cm box, draw lines vertically and horizontally to create smaller boxes. The children will need two copies of this grid.

2 Cut a small rectangle of poly board approximately 3cm x 2cm.

3 Using two similar colours, such as red and orange, or blue and green, dip the short edge of the poly board into the paint and print one 'cell' in the first box.

4 Print two cells in the next box, four cells in the box after that, continuing to double the quantity in each successive box until the boxes become very crowded. (The sequence is as follows: 1, 2, 4, 8, 16, 32, 64, 128.) When the boxes become very crowded, reverse the process on the second grid and reduce the number of cells by half each time. This represents the effect of antibiotics on bacteria, eventually destroying all bacteria cells by the end.

5 In an unused box draw a simple cartoon of an antibiotic. It should be drawn in the style of a superhero and can be a pill, or a bottle of medicine in a cape, for example.

6 Once dry, cut out each rectangle and stack them in the correct order. Place the image of the superhero antibiotic after the last cell division square and before the first cell reduction square. Add a front and back cover using a nice bright colour.

7 Staple all the rectangles together (teacher to do this) in one book and flick the pages to animate the cell division process.

Magnified Microbes

Magnified images of cells come in a variety of shapes, sizes and colours. Discuss the differences between bacteria and viral cells, yeast and fungi cells. This project looks at the variety of cells found in a range of life forms and substances. Sources can include blue cheese and penicillin which contain the same mould, bread products containing yeast, mushrooms that are fungi, viruses that cause a cold, and the difference between plant and animal cells.

Resources

- A2 and A4 white paper
- Acrylic or poster paints
- Laminator and pouches (optional)
- Cutting knife (teacher use only)
- Black sugar paper
- Glue sticks
- Coloured pencils
- Cardboard

Approach

1 Draw and paint a large version of the image (for example, a loaf of bread or a person sneezing) on the paper. Allow it to dry and then cut the painting out.
2 To make the magnifying glass, on a pre-cut circle of white paper draw and colour in the image of a cell. Draw a circle on black paper that is slightly larger than the white circle and cut it out. For the handle cut out a rectangle using the black paper.
3 To assemble, stick the magnifying glass on top of the main picture, and finally stick the disk onto the magnifying glass.
- Option: to use as a teaching aid, mount the work on stiff card and cut around the handle of the magnifying glass, so that each piece can be held up in the classroom.

Titles in this series:

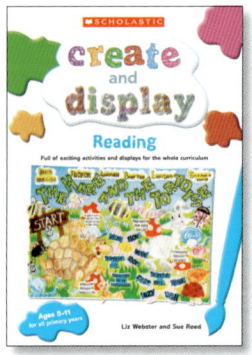

ISBN: 978-1-407-11915-1

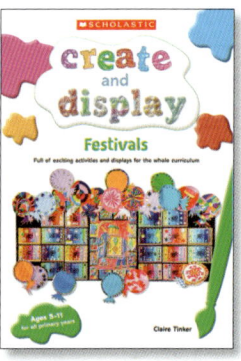

SBN: 978-1-407-11918-2

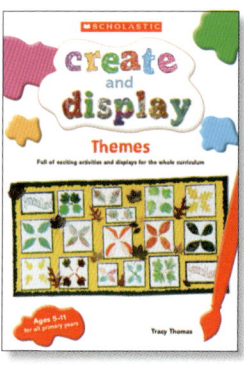

ISBN: 978-1-407-11916-8

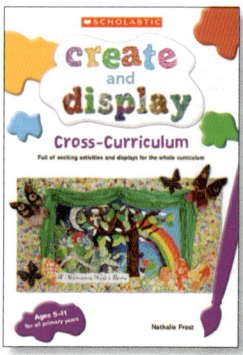

ISBN: 978-1-407-11917-5

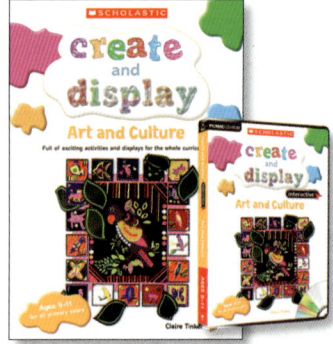

ISBN: 978-1-407-12527-5 (Book)
ISBN: 978-1-407-12533-6 (CD-ROM)

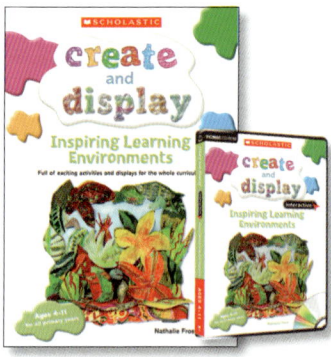

ISBN: 978-1-407-12526-8 (Book)
ISBN: 978-407-12532-9 (CD-ROM)

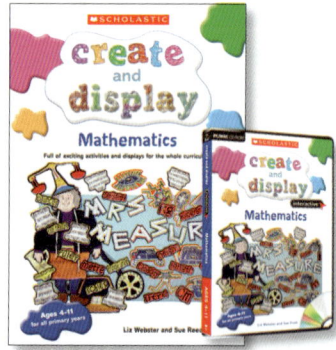

ISBN: 978-407-12525-1 (Book)
ISBN: 978-407-12531-2 (CD-ROM)

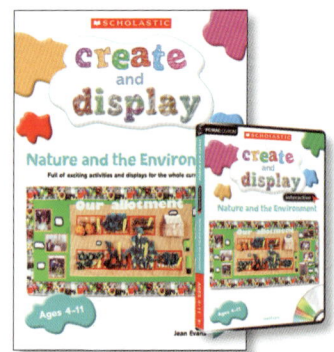

ISBN: 978-407-12528-2 (Book)
ISBN: 978-407-12534-3 (CD-ROM)

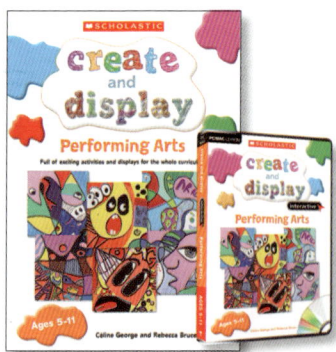

ISBN: 978-1-407-12530-5 (Book)
ISBN: 978-407-12536-7 (CD-ROM)

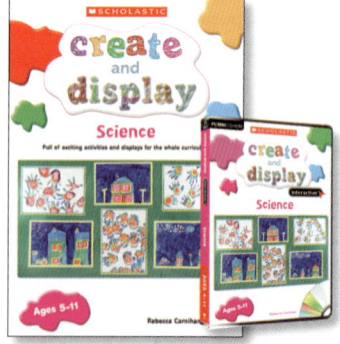

ISBN: 978-1-407-12529-9 (Book)
ISBN: 978-407-12535-0 (CD-ROM)

To find out more, call: **0845 603 9091**
or visit our website **www.scholastic.co.uk**